Unified Fracture Design

Bridging the Gap Between Theory and Practice

Unified Fracture Design
Bridging the Gap Between Theory and Practice

Michael Economides
Ronald Oligney
Peter Valkó

Orsa Press

Alvin, Texas

Unified Fracture Design

Orsa Press
P.O. Box 2569
Alvin, TX   77512

10 9 8 7 6 5 4 3 2 1

Library of Congress Cataloging-in-Publication Data

Economides, Michael J.
    Unified fracture design : bridging the gap between theory and practice / Michael Economides, Ronald Oligney, Peter Valkó
        p. cm.
    Includes bibliographical references and index.
    ISBN 0-9710427-0-5 (alk. paper)
    1. Oil wells—Hydraulic fracturing. I. Oligney, Ronald E.
II. Valkó, Peter, 1950- III. Title.

TN871.E335 2002
622'.3382—dc21   2002025793

Printed in the United Kingdom

Contents

CHAPTER 1

Hydraulic Fracturing for Production or Injection Enhancement 1

- FRACTURING AS COMPLETION OF CHOICE 1
- BASIC PRINCIPLES OF UNIFIED FRACTURE DESIGN 4
  - Fractured Well Performance 5
  - Sizing and Optimization 7
  - Fracture-to-Well Connectivity 8
- THE TIP SCREENOUT CONCEPT AND OTHER ISSUES IN HIGH PERMEABILITY FRACTURING 9
  - Tip Screenout Design 10
  - Net Pressure and Leakoff in the High Permeability Environment 11
  - Candidate Selection 12
- "BACK OF THE ENVELOPE" FRACTURE DESIGN 14
  - Design Logic 14
  - Fracture Design Spreadsheet 14

CHAPTER 2

How To Use This Book 17

- STRUCTURE OF THE BOOK 17
- WHICH SECTIONS ARE FOR YOU 18
  - Fracturing Crew 19

CHAPTER 3

Well Stimulation as a Means to Increase the Productivity Index 23

■ PRODUCTIVITY INDEX   23
■ THE WELL-FRACTURE-RESERVOIR SYSTEM   26
■ PROPPANT NUMBER   27
Well Performance for Low and Moderate Proppant Numbers   32
■ OPTIMUM FRACTURE CONDUCTIVITY   35
■ DESIGN LOGIC   38

CHAPTER 4

Fracturing Theory 39

■ LINEAR ELASTICITY AND FRACTURE MECHANICS   39
■ FRACTURING FLUID MECHANICS   42
■ LEAKOFF AND VOLUME BALANCE IN THE FRACTURE   46
Formal Material Balance: The Opening-Time Distribution Factor   46
Constant Width Approximation (Carter Equation II)   48
Power Law Approximation to Surface Growth   48
Detailed Leakoff Models   50

■ BASIC FRACTURE GEOMETRIES   50
Perkins-Kern Width Equation   51
Khristianovich-Zheltov-Geertsma-deKlerk Width Equation   53
Radial (Penny-shaped) Width Equation   54

CHAPTER 5

Fracturing of High Permeability Formations 57

■ THE EVOLUTION OF THE TECHNIQUE   57
■ HPF IN VIEW OF COMPETING TECHNOLOGIES   60
Gravel Pack   60
High-Rate Water Packs   62
■ PERFORMANCE OF FRACTURED HORIZONTAL WELLS
IN HIGH PERMEABILITY FORMATIONS   62

■ DISTINGUISHING FEATURES OF HPF   63
The Tip Screenout Concept   63
Net Pressure and Fluid Leakoff   68
Net Pressure, Closure Pressure, and Width in Soft Formations   66
Fracture Propagation   66
■ LEAKOFF MODELS FOR HPF   67
Fluid Leakoff and Spurt Loss as Material Properties: The Carter Leakoff
Model with Nolte's Power Law Assumption   67
Filter Cake Leakoff Model According to Mayerhofer, et al.   68
Polymer-Invaded Zone Leakoff Model of Fan and Economides   70
■ FRACTURING HIGH PERMEABILITY GAS CONDENSATE RESERVOIRS   72
Optimizing Fracture Geometry in Gas Condensate Reservoirs   74
■ EFFECT OF NON-DARCY FLOW IN THE FRACTURE   76
Definitions and Assumptions   77
Case Study for the Effect of Non-Darcy Flow   80

CHAPTER 6

Fracturing Materials   83

■ FRACTURING FLUIDS   84
■ FLUID ADDITIVES   85
■ PROPPANTS   87
Calculating Effective Closure Stress   88
■ FRACTURE CONDUCTIVITY AND MATERIALS SELECTION IN HPF   91
Fracture Width as a Design Variable   91
Proppant Selection   93
Fluid Selection   94

CHAPTER 7

Fracture Treatment Design   101

■ MICROFRACTURE TESTS   101
■ MINIFRACS   102
■ TREATMENT DESIGN BASED ON THE UNIFIED APPROACH   109
Pump Time   110
Proppant Schedule   113
Departure from the Theoretical Optimum   118
TSO Design   119

■ PUMPING A TSO TREATMENT 120
Swab Effect Example 121
Perforations for HPF 122

■ PRE-TREATMENT DIAGNOSTIC TESTS FOR HPF 122
Step-Rate Tests 123
Minifracs 125
Pressure Falloff Tests 126
Bottomhole Pressure Measurements 126

CHAPTER 8

Fracture Design and Complications 129

■ FRACTURE HEIGHT 129
Fracture Height Map 132
Practical Fracture Height Determination 133

■ TIP EFFECTS 134

■ NON-DARCY FLOW IN THE FRACTURE 135

■ COMPENSATING FOR FRACTURE FACE SKIN 136

■ EXAMPLES OF PRACTICAL FRACTURE DESIGN 137
A Typical Preliminary Design—Medium Permeability Formation: MPF01 137
Pushing the Limit—Medium Permeability Formation: MPF02 142
Proppant Embedment: MPF03 145
Fracture Design for High Permeability Formation: HPF01 148
Extreme High Permeability: HPF02 152
Low Permeability Fracturing: LPF01 157

■ SUMMARY 162

CHAPTER 9

Quality Control and Execution 163

■ FRACTURING EQUIPMENT 164

■ EQUIPMENT LIST 166
Water Transfer and Storage 166
Proppant Supply 167
Slurrification and Blending 168
Pumping 169
Monitoring and QA/QC 172
Miscellaneous 174

■ SPECIAL INSTRUCTIONS ON HOOK-UP 174
Spotting the Equipment 174

Fluid Supply-to-Blender   177
Proppant Supply   177
Frac Pumps   177
Manifold-to-Well   178
Monitoring/Control Equipment and Support Personnel   179
■ STANDARD FRACTURING QA PROCEDURES   180
■ FORCED CLOSURE   181
■ QUALITY CONTROL FOR HPF   183

CHAPTER 10

Treatment Evaluation   185

■ REAL-TIME ANALYSIS   185
■ HEIGHT CONTAINMENT   186
■ LOGGING METHODS AND TRACERS   188
■ A WORD ON FRACTURE MAPPING   189
■ WELL TESTING   190
■ EVALUATION OF HPF TREATMENTS—A UNIFIED APPROACH   193
Production Results   193
Evaluation of Real-Time HPF Treatment Data   194
Post-Treatment Well Tests in HPF   195
Validity of the Skin Concept in HPF   197
■ SLOPES ANALYSIS   197
Assumptions   198
Restricted Growth Theory   199
Slopes Analysis Algorithms   201

Appendices

A: NOMENCLATURE   207
B: GLOSSARY   211
C: BIBLIOGRAPHY   219
D: FRACTURE DESIGN SPREADSHEET   227
E: MINIFRAC SPREADSHEET   233
F: STANDARD PRACTICES AND QC FORMS   239
G: SAMPLE FRACTURE PROGRAM   251

Index   259

Preface

The purpose of writing this book is to establish a unified design methodology for hydraulic fracture treatments, a long established well stimulation activity in the petroleum and related industries. Few activities in the industry hold such potential to improve well performance both profitably and reliably.

The word "unified" has been selected deliberately to denote both the integration of all the highly diverse technological aspects of the process, but also to dispel the popular notion that there is one type of treatment that applies to low-permeability and another to high-permeability reservoirs. It is natural, even for experienced practitioners to think so because traditional targets have been low-permeability reservoirs while the fracturing of high-permeability formations has sprung from the gravel pack, sand control practice.

The key idea is that treatment sizes can be unified because they can be best characterized by the dimensionless Proppant Number, which determines the theoretically optimum fracture dimensions at which the maximum productivity or injectivity index can be obtained. Technical constraints should be satisfied in such a way that the design departs from the theoretical optimum only to the necessary extent. With this approach, difficult topics such as high- versus low-permeability fracturing, extensive height growth, non-Darcy flow, and proppant embedment are treated in a transparent and unified way, providing the engineer with a logical and coherent design procedure.

A design software package is included with the book.

The authors' backgrounds span the entire spectrum of technical, research, development, and field applications in practically all geographic and reservoir type settings. It is their desire that this book finds its appropriate place in everyday practice.

Hydraulic Fracturing for Production or Injection Enhancement

FRACTURING AS COMPLETION OF CHOICE

This book has the ambition to do something that has not been done properly before: to unite the gap between theory and practice in what is arguably the most common stimulation/well completion technique in petroleum production. Even more important, the book takes a new and ascendant position on the most critical link in the sequence of events in this type of well stimulation—the sizing and the design of hydraulic fracture treatments.

Fracturing was first employed to improve production from marginal wells in Kansas in the late 1940s (Figure 1-1). Following an explosion of the practice in the mid-1950s and a considerable surge in the mid-1980s, massive hydraulic fracturing (MHF) grew to become a dominant completion technique, primarily for low permeability reservoirs in North America. By 1993, 40 percent of new oil wells and 70 percent of gas wells in the United States were fracture treated.

With improved modern fracturing capabilities and the advent of high permeability fracturing (HPF), which in the vernacular has been referred to as "frac & pack" or variants, fracturing has expanded further to become the completion of choice for all types of wells in the United States, but particularly natural gas wells (see Figure 1-2).

1

FIGURE 1-1. An early hydraulic fracture treatment, circa. 1949. (*Source:* Halliburton.)

The tremendous advantage in fracturing most wells is now largely accepted. Even near water or gas contacts, considered the bane of fracturing, HPF is finding application because it offers controlled fracture extent and limits drawdown (Mullen et al., 1996; Martins et al., 1992). The rapid ascent of high permeability fracturing from a few isolated treatments before 1993 (Martins et al., 1992; Grubert, 1991; Ayoub et al., 1992) to some 300 treatments per year in the United States by 1996 (Tiner et al., 1996) was the start to HPF becoming a dominating optimization tool for integrated well completion and production. Today, it is established as one of the major recent developments in petroleum production.

The philosophy of this book hinges on the overriding commonality in fracture design that transcends the value of the reservoir permeability. There is a strong theoretical foundation to this approach, which will be outlined in this book. Hence the title, *Unified Fracture Design,* which suggests the connection between theory and practice, but also that the design process cuts across all petroleum reservoirs—low permeability to high permeability, hard rock to soft rock. And indeed, it is common to all.

There is substantial room for additional growth of hydraulic fracturing in the worldwide petroleum industry, as well as other industries. It is estimated that hydraulic fracturing may add several hundred thousand barrels per day from existing wells in a number of countries.

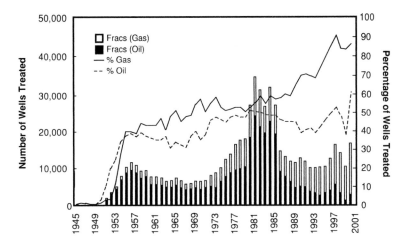

FIGURE 1-2. Fracturing as "completion of choice" in U.S. oil and gas wells. (*Source:* Schlumberger.)

This can be accomplished if the process is undertaken in a serious and concerted way so that the economy of scale influences the cost of the treatments and, hence, the overall economics.

There are two frequently encountered impediments to substantial applications of hydraulic fracturing:

1. A widespread misunderstanding that the process is still only for low permeability reservoirs (e.g., less than 1 md), or that it is the last refuge for enhancing well production or injection performance, to be tried only if everything else fails. The latter carries along with it an often unjustifiable phobia that hydraulic fracturing is dangerous, that it accelerates the onset of water production, that it increases the water cut or affects zonal isolation, and so on. The more serious associated problem is that using fracturing as a last, at times desperate, resort implies unplanned stimulation that may suffer from a number of problems (such as well deviation and inadequate perforating), which, in turn, may almost guarantee disappointing results. A final related problem is the notion that high permeability fracturing applies only to those reservoirs that need sand production control. This is clearly not the case, and reservoirs with permeabilities of several hundred millidarcies are now routinely fractured.

2. At times, engineers in various companies outside of North America may actually try fracturing, but a treatment is done so rarely

and so haphazardly that it is bound to be expensive, such that the cost cannot be justified even if the incremental production is substantial. Hydraulic fracturing is a massive operation with a very large complement of equipment, complicated and demanding fluids and proppants, and a wide spectrum of ancillary and people-intensive engineering and operational demands. Costs assigned to individual, isolated jobs—e.g., one or two treatments carried out every three to six months—are essentially prohibitive. Coupled with an occasional job failure, sketchy and spotty application of hydraulic fracturing is almost assured of economic failure and the dampening of any desire to apply it further.

Virtually no petroleum operation carries such a differential price tag among areas where it is applied in a widespread and massive way, such as North America and offshore in the North Sea, and elsewhere. In North America, over 60 percent of all oil wells and 85 percent of all gas wells are hydraulically fractured, and the percentages are still increasing. Yet, consider this: a 100-ton proppant treatment in the United States, at the time of this writing, costs less than $100,000. Exactly the same treatment, with the same equipment and the same service company, for example in Venezuela or Oman, is likely to cost at least $1 million, and it can cost as much as $2 million.

At the same time, virtually no other petroleum technology carries a larger incremental asset. The hundreds-of-thousands to millions of barrels per day of worldwide production increase that we project assumes that the percentage of existing wells being hydraulically fractured approaches that of oil wells in the United States (60 percent), and the incremental production realized from each well is just 25 percent over the pre-treatment state. The latter implies the very modest assumptions that all existing wells continue to produce, and that fracturing would result in a very achievable average "skin" equal to –2. In fact, the incremental production capacity from a massive stimulation campaign with adequate equipment and well-trained people is likely to be much higher.

BASIC PRINCIPLES OF UNIFIED FRACTURE DESIGN

Hydraulic fracturing entails injecting fluids in an underground formation at a pressure that is high enough to induce a parting of the formation.

Granulated materials—called "proppants," which range from natural sands to rather expensive synthetic materials—are pumped into the created fracture as a slurry. They hold open, or "prop," the created fracture after the injection pressure used to generate the fracture has been relieved.

The fracture, filled with proppant, creates a narrow but very conductive flow path toward the wellbore. This flow path has a very large permeability, frequently five to six orders of magnitude larger than the reservoir permeability. It is most often narrow in one horizontal direction, but can be quite long in the other horizontal direction and can cover a significant height. Typical intended propped widths in low permeability reservoirs are on the order of 0.25 cm (0.1 in.), while the length can be several hundred meters. In high permeability reservoirs, the targeted fracture width (deliberately affected by the design and execution) is much larger, perhaps as high as 5 cm (2 in.), while the length might be as short as 10 meters (30 ft).

In almost all cases, an overwhelming part of the production comes into the wellbore through the fracture; therefore, the original near-wellbore damage is "bypassed," and the pre-treatment skin does not affect the post-treatment well performance.

Fractured Well Performance

The performance of a fractured well can be described in many ways. One common way is to forecast the production of oil, gas, and even water as a function of time elapsed after the fracturing treatment. However, post-treatment production is influenced by many decisions that are not crucial to the treatment design itself. The producing well pressure, for example, may or may not be the same as the pre-treatment pressure, and may or may not be held constant over time. Even if, for the sake of evaluation, an attempt is made to set all well operating parameters the same before and after the treatment, comparison over time is still obfuscated by the accelerated nature of reservoir depletion in the presence of a hydraulic fracture.

Thus, in a preliminary sizing and optimization phase, it is imperative to use a simple performance index that describes the expected and actual improvement in well performance due to the treatment.

In unified fracture design, we consider a very simple and straightforward performance indicator: the *pseudo-steady state productivity index*. The improvement in this variable describes the actual effect of the propped fracture on well performance. Realizing the maximum possible pseudo-steady state productivity index, for all practical purposes, means that the fracture will not under-perform any other possible realization of the same propped volume, even if the well produces for a considerable time period in the so-called "transient" regime. While this statement might not appear plausible at first, the experienced production engineer will understand it by thinking of the transient flow period as a continuous increase in *drainage area* in which the pseudo-steady state has already been established. Considerable cumulative production can only come from a large drained area, and hence that pseudo-steady state productivity index must be maximized, which corresponds to the finally formed drainage area.

Fracture length and *dimensionless fracture conductivity* are the two primary variables that control the productivity index of a fractured well. Dimensionless fracture conductivity is a measure of the relative ease with which produced fluids flow inside the fracture compared to the ability of the formation to feed fluids *into* the fracture. It is calculated as the product of fracture permeability and fracture width, divided by the product of reservoir permeability and fracture (by convention, half-) length.

In low permeability reservoirs, the fracture conductivity is *de facto* large, even if only a narrow propped fracture has been created and a long fracture length is needed. A post-treatment skin can be as negative as −7, leading to several folds-of-increase in well performance compared to the unstimulated well.

For high permeability reservoirs, a large fracture width is essential for adequate fracture performance. Over the last several years, a technique known as tip screenout (TSO) has been developed, which allows us to deliberately arrest the lateral growth of a hydraulic fracture and subsequently inflate its width, exactly to affect a larger conductivity.

For a fixed volume of proppant placed in the formation, a well will deliver the maximum production or injection rate when the dimensionless fracture conductivity is near unity. In other words, a dimensionless fracture conductivity around one (or more precisely, 1.6, as shown in Chapter 3) is the physical optimum, at least for treatments not involving extremely large quantities of proppant. Larger values of the dimensionless fracture conductivity would mean relatively shorter-than-optimum fracture lengths and, thus, the flow from the reservoir into the fracture would be unnecessarily restricted. Dimensionless fracture conductivity values *smaller* than unity would mean less-than-optimum fracture *width,* rendering the fracture as a bottleneck to optimum production.

There are a number of secondary issues that complicate the picture—early time transient flow regime, influence of reservoir boundaries, non-Darcy flow effects, and proppant embedment, just to mention a few. Nevertheless, these effects can be correctly taken into account only if the role of dimensionless fracture conductivity is understood.

It is possible that in certain theaters of operation the practical optimum may be different than the physical optimum. In some cases, the theoretically indicated fracture geometry may be difficult to achieve because of physical limitations imposed either by the available equipment, limits in the fracturing materials, or the mechanical properties of the rock to be fractured. However, aiming to maximize the well productivity or injectivity is an appropriate first step in the fracture design.

Sizing and Optimization

The term "optimum" as used above means the maximization of a well's productivity, within the constraint of a certain treatment size. Hence, a decision on treatment size should actually precede (or go hand-in-hand with) an optimization based on the dimensionless fracture conductivity criterion.

For a long time, practitioners considered fracture half-length as a convenient variable to characterize the size of the created fracture. That tradition emerged because it was not possible to independently change fracture length and width, and because length had a primary effect on productivity in low permeability formations. In unified fracture design, where both low and high permeability formations are considered, the best single variable to characterize the size of a created fracture is the *volume of proppant* placed in the productive horizon, or "pay."

Obviously, the total volume of proppant placed in the pay interval is always less than the total proppant injected. From a practical point of view, treatment sizing means deciding how much proppant to inject. When sizing the treatment, an engineer must be aware that increasing the injected amount of proppant by a certain quantity x will not necessarily increase the amount of proppant reaching the pay layer by the same quantity, x. We will refer to the ratio of the two proppant volumes (i.e., the volume of proppant placed in the pay interval divided by the total volume injected into the well) as the *volumetric proppant efficiency.*

By far the most critical factor in determining volumetric proppant efficiency is the ratio of created fracture height to the net pay thickness.

Extensive height growth limits the volumetric proppant efficiency, and is something that we generally try to avoid. (The possibility of intersecting a nearby water table is another important reason to avoid excessive height growth.)

Actual selection of the amount of proppant indicated for injection is primarily based on economics, the most commonly used criterion being the net present value (NPV). As with most engineering activities, costs increase almost linearly with the size of the treatment, but after a certain point, the revenues increase only marginally. Thus, there is an optimum treatment size, the point at which the NPV of incremental revenue, balanced against treatment costs, is a maximum.

The optimum size can be determined if some method is available to predict the maximum possible productivity increase achievable with a certain amount of proppant. Unified fracture design makes extensive use of this fact, given that the maximum achievable productivity increase is already determined by the volume of proppant in the pay. Many of the operational details are subsumed by the basic decision on treatment size, making possible a simple yet robust design process.

Therefore, we employ the concept of "volume of proppant reaching the pay" or simply "propped volume in the pay" as the key decision variable in the sizing phase of the unified fracture design procedure. To handle it correctly, the amount of proppant indicated for injection and the volumetric proppant efficiency must be determined.

Fracture-to-Well Connectivity

While the maximum achievable improvement of productivity is determined by the propped volume in the pay, several additional conditions must be satisfied *en route* to a fracture that actually realizes this potential improvement. One of the crucial factors is to establish an optimum compromise between the length and width (or to depart from the optimum only as much as necessary, if required by operational constraints). As previously explained, the optimum dimensionless fracture conductivity is the variable that helps us to find the right compromise. However, another condition is equally important. It is related to the connectivity of the fracture to the well.

A reservoir at depth is under a state of stress that can be characterized by three principal stresses: one vertical, which in almost all cases of deep reservoirs (depth greater than 500 meters, 1500 ft) is the largest of the three, and two horizontal, one minimum and one maximum. A hydraulic fracture will be normal to the smallest stress, leading to vertical hydraulic fractures in almost all petroleum production applications. The azimuth of these fractures is pre-ordained by the natural state of earth stresses. As such, deviated or horizontal wells that are to be fractured should be drilled in an orientation that agrees with this azimuth. Vertical wells, of course, naturally coincide with the fracture plane.

If the well azimuth does not coincide with the fracture plane, the fracture is likely to initiate in one plane and then twist, causing considerable "tortuosity" *en route* to its final azimuth—normal to the minimum stress direction. Vertical wells with vertical fractures or perfectly horizontal wells drilled deliberately along the expected fracture plane result in the best aligned well-fracture systems. Other well-fracture configurations are subject to "choke effects," unnecessarily decreasing the productivity of the fractured well. Perforations and their orientation may also be a source of problems during the execution of a treatment, including multiple fracture initiations and premature screenouts caused by tortuosity effects.

The dimensionless fracture conductivity in low permeability reservoirs is naturally high, so the impact of choke effects from the phenomena described above is generally minimized; to avoid tortuosity, point source fracturing is frequently employed.

Fracture-to-well connectivity is considered today as a critical point in the success of high permeability fracturing, often dictating the well azimuth (e.g., drilling S-shape vertical wells) or indicating horizontal wells drilled longitudinal to the fracture direction. Perforating is being revisited, and alternatives, such as hydro-jetting of slots, are considered by the most advanced practitioners. While some models incorporate complex well-fracture geometries with choke and other effects, the many uncertainties prevent us from predicting performance. Rather, we are limited to *explain* the performance once post-treatment well test and production information become available. In the design phase, we try to make decisions that minimize the likelihood of such unnecessary reductions in productivity.

THE TIP SCREENOUT CONCEPT AND OTHER ISSUES IN HIGH PERMEABILITY FRACTURING

Because high permeability fracturing has the most fertile possibility for expansion in petroleum operations worldwide, key issues for this type of well completion are described below. The purpose is to identify those features that distinguish high permeability fracturing from conventional hydraulic fracturing.

Tip Screenout Design

The critical elements of high permeability fracturing treatment design, execution and treatment behavioral interpretation are substantially different than for conventional fracturing treatments. In particular, HPF relies on a carefully timed "tip screenout," or TSO, to limit fracture growth and allow for fracture inflation and packing. The TSO occurs when sufficient proppant has concentrated at the leading edge of the fracture to prevent further fracture extension. Once the fracture growth has been arrested (and assuming the pump rate is larger than the rate of leakoff to the formation), continued pumping will inflate the fracture, i.e., increase the fracture width. Tip screenout and fracture inflation should be accompanied by an increase in net fracturing pressure. Thus, the treatment can be conceptualized in two distinct stages: fracture creation (equivalent to conventional designs) and fracture inflation/packing (after tip screenout).

Creation of the fracture and arrest of its growth (i.e., the tip screenout) is accomplished by injecting a relatively small "pad" of clean fluid (no sand) followed by a "slurry" containing 1–4 lbs of sand per gallon of fluid (ppg). Once the fracture growth has been arrested, further injection builds fracture width and allows injection of a high-concentration slurry (e.g., 10–16 ppg). Final *areal proppant concentrations* of 20 lb_m/sq ft are possible. A usual practice is to retard the injection rate near the end of the treatment (coincidental with opening the annulus to flow) to dehydrate and pack the fracture near the well. Rate reductions may also be used to force a tip screenout in cases where no TSO event is observed on the downhole pressure record.

Frequent field experience suggests that the tip screenout can be difficult to model, affect, or even detect. There are many reasons for this, including a tendency toward overly conservative design models (resulting in no TSO), partial or multiple tip screenout events, and inadequate pressure monitoring practices.

Accurate bottomhole measurements are imperative for meaningful treatment evaluation and diagnosis. Calculated bottomhole pressures are unreliable because of the sizeable and complex friction pressure effects associated with pumping high proppant slurry concentrations through small diameter tubulars and service tool crossovers.

Surface data may indicate that a TSO event has occurred when the bottomhole data shows no evidence, and vice versa.

Net Pressure and Leakoff in the High Permeability Environment

The entire HPF process is dominated by *net pressure* and *fluid leakoff* considerations. First, high permeability formations are typically soft and exhibit low elastic modulus values, and second, the fluid volumes are relatively small and leakoff rates high (high permeability, compressible reservoir fluids and non-wall building fracturing fluids). While traditional practices applicable to design, execution, and evaluation in hydraulic fracturing continue to be used in HPF, these are frequently not sufficient.

Net Pressure

Net pressure is the difference between the pressure at any point in the fracture and the pressure at which the fracture will close. This definition implies the existence of a unique *closure pressure*. Whether the closure pressure is a constant property of the formation or depends heavily on the pore pressure (or rather on the disturbance of the pore pressure relative to the long-term steady value) is an open question.

In high-permeability, soft formations it is difficult (if not impossible) to suggest a simple recipe to determine the closure pressure as classically derived from shut-in pressure decline curves. Furthermore, because of the low elastic modulus values, even small induced uncertainties in the net pressure are amplified into large uncertainties in the calculated fracture width.

Fracture propagation, the availability of sophisticated 3D models notwithstanding, is a very complex process and difficult to describe, even in the best of cases, because of the large number and often competing physical phenomena. The physics of fracture propagation in soft rock is even more complex, but it is reasonably expected to involve incremental energy dissipation and more severe tip effects when compared to hard rock fracturing. Again, because of the low modulus values, an inability to predict net pressure behavior may lead to a significant departure between predicted and actual treatment performance. Ultimately, the classic fracture propagation models may not reflect even the main features of the propagation process in high permeability rocks.

It is common practice for some practitioners to "predict" fracture propagation and net pressure features *ex post facto* using a computer fracture simulator. The tendency toward substituting clear models and physical assumptions with "knobs"—e.g., arbitrary stress barriers, friction changes (attributed to erosion if decreasing and sand resistance if increasing) and less than well understood properties of the formation expressed as dimensionless factors—does not help to clarify the issue. Other techniques are warranted and several are under development.

Leakoff

Considerable effort has been expended on laboratory investigation of the fluid leakoff process for high permeability cores. The results raise some questions about how effectively fluid leakoff can be limited by filtercake formation. In all cases, but especially in high permeability formations, the quality of the fracturing fluid is only one of the factors that influence leakoff, and often not the determining one. Transient fluid flow in the formation might have an equal or even larger impact. Transient flow cannot be understood by simply fitting an empirical equation to laboratory data. The use of models based on solutions to the fluid flow in porous media is an unavoidable step, and one that has already been taken by many.

Candidate Selection

The utility of high permeability fracturing extends beyond the obvious productivity benefits associated with bypassing near-well damage to include *sand control*. However, in HPF the issue is not mere sand control, which implies most often mechanical retention of migrating sand particles (and plugging), but rather sand *deconsolidation* control.

Increasingly, wellbore stability should be viewed in a holistic approach with horizontal wells and hydraulic fracture treatments. Proactive well completion strategies are critical to wellbore stability and sand-production control to reduce pressure drawdown while obtaining economically attractive rates. Reservoir candidate recognition for the correct well configurations is the key element. Necessary steps in candidate selection include appropriate reservoir engineering, formation characterization, wellbore stability calculations, and the combining of production forecasts with assessments of sand-production potential.

Complex Well-Fracture Configurations

Vertical wells are not the only candidates for hydraulic fracturing. Figure 1-3 shows some basic single-fracture configurations for vertical and horizontal wells. Horizontal wells that employ conventional or especially high permeability fracturing with the well drilled in the expected fracture azimuth (accepting a longitudinal fracture) appear to have, at least conceptually, a very promising prospect as discussed in Chapter 5. However, a horizontal well intended for a longitudinal fracture configuration would have to be drilled along the maximum horizontal stress. And this, in addition to well-understood drilling problems, may contribute to long-term formation stability problems.

Figure 1-4 illustrates two multi-fracture configurations. A rather sophisticated conceptual configuration would involve the combination of HPF with multiple-fractured vertical branches emanating from a horizontal "mother" well drilled above the producing formation. Of course, horizontal wells, being normal to the vertical stress, are generally more prone to wellbore stability problems. Such a configuration would allow for placement of the horizontal borehole in a competent, non-producing interval. There are other advantages to fracture treating a vertical section over a highly deviated or horizontal section: multiple starter fractures, fracture turning, and tortuosity problems are avoided; convergence-flow skins (choke effects) are much less of a concern; and the perforating strategy is simplified.

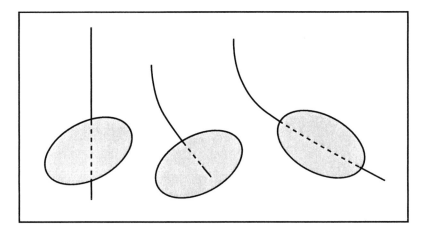

FIGURE 1-3. Single-fracture configurations for vertical and horizontal wells.

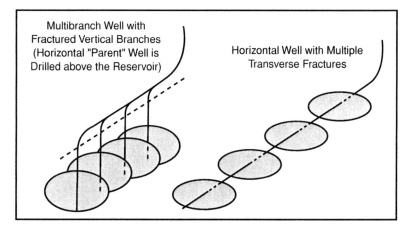

FIGURE 1-4. Multibranched, multiple-fracture configurations for horizontal wells.

"BACK OF THE ENVELOPE" FRACTURE DESIGN

Design Logic

In unified fracture design, we consider treatment size, specifically propped volume in the pay, as the primary decision variable. Once the basic decision on size is made, the optimum length and width are determined. These parameters are then revised in view of the technical constraints, and the target dimensions of the created fracture are set. A preliminary injection schedule is calculated that realizes the target dimensions and assures uniform placement of the indicated amount of proppant. If the optimum placement cannot be realized by traditional means, a TSO treatment is indicated. Even if the injected amount of proppant is already fixed, the volumetric proppant efficiency may change during the design process. It is extremely important that the basic decisions be made in an iterative manner, but without going into unnecessary details of fracture mechanics, fluid rheology, or reservoir engineering.

Fracture Design Spreadsheet

A simple spreadsheet, based on a transparent design logic, is an ideal tool to make preliminary design decisions and a primary evaluation of the executed treatment. The CD attached to the back cover of the book contains such a spreadsheet, named HF2D. The HF2D Excel spreadsheet is a fast 2D software package for the design of traditional (moderate permeability and hard rock) and frac & pack (higher permeability and soft rock) fracture treatments.

Readers are strongly encouraged to use the spreadsheet while reading the book. By modifying various input parameters, an intuitive feel for their relative importance in treatment design and final fractured well performance can be rapidly acquired, an important but uncommon prospect in the era of complex 3D fracture simulators. The spreadsheet will help readers make the most important decisions and be aware of their consequences.

The attached spreadsheet is not necessarily intended as a substitute for more sophisticated software tools, but the rapid "back of the envelope" calculations that it affords can provide substantive fracture designs. In many cases, by virtue of restricting the analysis to important first-order considerations, the spreadsheet results are more robust than those provided by highly involved 3D fracture simulators. It is suggested that readers run parallel cases with one or more 3D simulators, if available, as an interesting exercise.

How To Use This Book

The purpose of this book is to transfer hydraulic fracturing technology and, especially, facilitate its execution. The various chapters supply information on candidate recognition, fracture treatment design, execution and evaluation, materials selection, quality control, and equipment specifications.

While the book includes late developments from some of the most respected practitioners of hydraulic fracturing in the world—genuine state-of-the-art technology—the entry point is deliberately low. That is, the book can also serve as a very useful primer for those being exposed to fracturing technology for the first time.

STRUCTURE OF THE BOOK

Chapters 1 through 10 provide a detailed narrative of the most important aspects across the spectrum of hydraulic fracturing activities.

Appendices A through G are reference material, including a glossary of fracturing terms; an extensive bibliography; data requirements and user instructions for the included design software; standard quality control practices and forms; and example fracturing procedures.

The CD attached to the back cover of the book contains two spreadsheets:

1. The HF2D Excel spreadsheet is a fast 2D software package for the design of traditional (moderate permeability and hard rock) and frac & pack (higher permeability and soft rock) fracture treatments.
2. The MF Excel spreadsheet is a minifrac (calibration test) evaluation package. Its main purpose is to extract the leakoff coefficient from pressure fall-off data.

Two industry-leading references are strongly recommended as addenda to this book:

▪ *Hydraulic Fracture Mechanics,* by Peter Valkó and Michael Economides, addresses the theoretical background of this seminal technology. It provides a fundamental treatment of basic phenomena such as elasticity, stress distribution, fluid flow, and the dynamics of the rupture process. Contemporary design and analysis techniques are derived and improved using a comprehensive and unified approach.

▪ *Stimulation Engineering Handbook,* by John Ely, aptly covers many issues of fracture treatment implementation and quality control. This is a very hands-on book, intended to drive execution performance and quality control.

Other reference books that contain abundant information by dozens of experts in the field include *Petroleum Well Construction,* edited by Michael Economides, Larry Watters, and Shari Dunn-Norman; *Reservoir Stimulation, Third Edition,* by Michael Economides and Ken Nolte; and the somewhat dated but classic volume, *SPE Monograph No. 12: Advances in Hydraulic Fracturing,* edited by John Gidley, Steve Holditch, Dale Nierode, and Ralph Veatch. While these books provide historical perspective as well as in-depth discussion and opinions (some controversial) on various details of the fracturing process, they are not recommended for a first reading because of the highly technical language and compartmentalized style of presentation.

WHICH SECTIONS ARE FOR YOU

Which sections of the book that you will use—whether it's a quick review of the introductory material or a check of the glossary, reading the chapter on fracturing fluids, only, or hands-on use of the design theory and software—depends on your role in the fracturing operation.

Neither this book nor any other technology transfer mechanism is useful apart from capable people. The following key personnel comprise the fracturing team and the targeted readership of this book.

Fracturing Crew

A fracturing crew is the absolute minimum and basic unit required for a fracturing treatment. The crew may consist of anywhere from 7 to 15 people, depending on the number of pumping units and the monitoring capability on location. Many of these people are trained to do multiple jobs, such as driving trucks, hooking up equipment, and installing and maintaining the monitoring instruments.

In addition to being trained on each piece of equipment that they will operate, each member of the fracturing crew should be conversant with the material in Chapter 10, On-Site Quality Control, and the accompanying Appendix F, Standard Practices and QC Forms.

The key people in any fracturing operation, in order of critical importance, are:

Frac-Crew Chief—Sometimes known as the *field engineer,* this is the person responsible on-site for proper execution of the job. He is a highly experienced person, either an engineer that has reverted into a field service manager position, or a highly gifted operator who has been promoted to the job. The crew chief directs fracturing operations from the monitoring van ("frac van") and has complete responsibility for the operation, including safety. He communicates constantly by two-way radio with all pumping, blender, and proppant storage operators. He is certified to operate high pressure equipment. He understands the fracture design and is responsible for its implementation. He has complete authority to continue or shut down a job. (Note that while the pronoun "he" is used for clarity, there are several highly capable women currently practicing as fracturing engineers.)

This is not a job that can be learned gradually in a start-up operation. This individual must be identified through a careful search among qualified candidates. Extensive and relevant hands-on experience in fracture execution is a must. The frac-crew chief should be highly conversant with *Unified Fracture Design* in its entirety.

Desk Engineer—The desk engineer concept is practiced by many companies, within and external to the petroleum industry. Simply put, the fracturing service company places one of its full-time staff permanently on location in each client producing company. The client is responsible to furnish a space (desk) at which the external employee (engineer) can sit and work, giving rise to the term *desk engineer*. This constant accessibility and the cross-pollination of needs (producing company) and capabilities (service company) can dramatically improve the range and success of application of a technology, and could be especially important for the rapid and necessarily massive introduction of hydraulic fracturing in a new operating area or country.

This individual will have the same aptitude as the frac-crew chief, but typically with somewhat less experience. Like the frac-crew chief, the desk engineer should become highly conversant with the entire fracturing book.

QA/QC Chemist—Any fracturing operation requires a chemist who is well versed in the chemistry and physics (rheology) of fracturing fluids and additives. This person operates a specially outfitted laboratory. The laboratory includes, in addition to all basic implements and working spaces (e.g., hoods), a Fann 50 high-pressure/high-temperature viscometer and possibly a fluid shear-history simulator. The chemist should have a background in polymer chemistry, or at least a good understanding of the subject matter, and should be trained in detecting the quality of proppant (visually, with a 100-magnification microscope).

The chemist is the field quality assurance/quality control (QA/QC) officer. Prior to the fracture treatment, he inspects the make-up water, fluid additives, and proppant to ensure that they are appropriate and that they are of high quality. During the treatment, he makes sure that the fracturing materials are blended in the correct proportions and at the proper time (e.g., in the case of delayed crosslinkers). He continues to spot check and approve the proppant quality in real-time for the duration of the treatment.

It is almost entirely the responsibility of the QC/QA chemist to understand Chapter 6 and Chapter 9 of this book, as well as Appendix F, and to revise them for company-specific needs. In addition, this person should fully digest the *Stimulation Engineering Handbook*.

Fracture Design Engineer—As the title suggests, this individual is responsible for design of the fracturing treatment. The fracture design engineer must master the basics of hydraulic fracturing, as included in Chapters 4 through 9, and should be proficient to run the included fracture design software. Depending on the magnitude of the fracturing activity, there could be several people trained to perform this task. In small operations, the same person may double-up as the field engineer that performs real-time analysis of the treatment from the frac van (Chapter 10).

The fracture design engineer must have an engineering background, preferably petroleum engineering, and be dedicated to study the subtle and sometimes complex aspects of fracture design. Experience in the industry is desirable, but not necessary. With proper training, a gifted person can start functioning properly after several jobs. Ultimately, the fracturing engineer should be broadly conversant in fracture execution, fracturing fluid chemistry, and well completions. He should be able to make critical use of the additional literature recommended above.

Well Stimulation as a Means to Increase the Productivity Index

The primary goal of well stimulation is to increase the productivity of a well by removing damage in the vicinity of the wellbore or by superimposing a highly conductive structure onto the formation. Commonly used stimulation techniques include hydraulic fracturing, frac & pack, carbonate and sandstone matrix acidizing, and fracture acidizing. Any of these stimulation techniques can be expected to generate some increase in the productivity index, which, in turn, can be used either to increase the production rate or decrease the pressure drawdown. There is no need to explain the benefits of increasing the production rate. The benefits of decreased pressure drawdown are less obvious, but include minimizing sand production and water coning and/or shifting the phase equilibrium in the near-well zone to reduce condensate formation. Injection wells also benefit from stimulation in a similar manner.

To understand how stimulation increases productivity, basic production and reservoir engineering concepts are presented below.

PRODUCTIVITY INDEX

In discussing the productivity of a specific well, we think of a linear relation between the production rate and the driving force (pressure drawdown),

$$q = J\Delta p \qquad (3\text{-}1)$$

where the proportionality "constant" J is called the productivity index (PI). During its lifespan, a well is subject to several changes in flow conditions, but the two most important idealizations are constant production rate,

$$\Delta p = \frac{\alpha_1 B q \mu}{2\pi k h} p_D \qquad (3\text{-}2)$$

and constant drawdown pressure,

$$q = \frac{2\pi k h \Delta p}{\alpha_1 B \mu} q_D \qquad (3\text{-}3)$$

where k is the formation permeability, h is the pay thickness, B is the formation volume factor, μ is the fluid viscosity, and α_1 is a conversion constant (equal to 1 for a coherent system). Either the production rate (q) or the drawdown (Δp) are specified, and therefore used to define the dimensionless variables. Table 3-1 lists some of the well-known solutions to the radial diffusivity equation.

Because of the radial nature of flow, most of the pressure drop occurs near the wellbore, and any damage in this region significantly increases the pressure loss. The impact of damage near the well can be represented by the skin factor, s, added to the dimensionless pressure in the expression of the PI:

$$J = \frac{2\pi k h}{B\mu(p_D + s)} \qquad (3\text{-}4)$$

The skin is another idealization, capturing the most important aspect of near-wellbore damage: the additional pressure loss caused by the

TABLE 3-1. Flow into an Undamaged Vertical Well

Flow Regime	Δp	$p_D \ (\approx 1/q_D)$
Transient (infinite acting reservoir)	$p_i - p_{wf}$	$p_D = -\frac{1}{2} Ei\left(-\frac{1}{4t_D}\right)$, where $t_D = \frac{kt}{\phi \mu c_t r_w^2}$
Steady state	$p_e - p_{wf}$	$p_D = \ln(r_e/r_w)$
Pseudo-steady state	$\bar{p} - p_{wf}$	$p_D = \ln(0.472 r_e/r_w)$

damage is proportional to the production rate. Even with best drilling and completion practices, some kind of near-well damage is present in most cases. The skin can be considered as the measure of the "goodness" of a well. Other mechanical factors, not caused by damage *per se* may add to the skin effect. These may include bad perforations, partial well penetration, or undersized well completion equipment, and so on. If the well is damaged (or its productivity is less than the ideal reference value for any other reason), the skin factor is positive.

Well stimulation increases the productivity index. It is reasonable to look at any type of stimulation as an operation to reduce the skin factor. With the generalization to negative values of skin factor, even such stimulation treatments—which not only remove damage but also superimpose some new or improved conductivity paths—can be put into this framework. In the latter case, it is more correct to speak about *pseudo-skin factor,* indicating that stimulation causes some structural changes in the fluid flow path as well as removing damage.

As we explained in Chapter 1, crucial from the fracture design viewpoint is the pseudo-steady state productivity index:

$$J = \frac{q}{\bar{p} - p_{wf}} = \frac{2\pi kh}{\alpha_1 B\mu} J_D \qquad (3\text{-}5)$$

where J_D is called the dimensionless productivity index.

For a well located in the center of a circular drainage area, the dimensionless productivity index in pseudo-steady state reduces to

$$J_D = \frac{1}{\ln\left[\dfrac{0.472 r_e}{r_w}\right] + s} \qquad (3\text{-}6)$$

In the case of a propped fracture, there are several ways to incorporate the stimulation effect into the productivity index. One can use the pseudo-skin concept,

$$J_D = \frac{1}{\ln\left[\dfrac{0.472 r_e}{r_w}\right] + s_f} \qquad (3\text{-}7)$$

or the equivalent wellbore radius concept,

$$J_D = \frac{1}{\ln\left[\dfrac{0.472 r_e}{r_w'}\right]} \qquad (3\text{-}8)$$

or one can just provide the dimensionless productivity index as a function of the fracture parameters,

$$J_D = \frac{\textit{function of drainage-volume geometry}}{\textit{and fracture parameters}} \qquad (3\text{-}9)$$

All three options give exactly the same results (if done in coherent terms). The last option is the most general and convenient, especially if we wish to consider fractured wells in more general drainage areas (not necessarily circular).

Many authors have provided charts and correlations in one form or another to handle special geometries and reservoir types. Unfortunately, most of the results are less obvious or difficult to apply in higher permeability cases. Even for the simplest possible case, a vertical fracture intersecting a vertical well, there are quite large discrepancies (see, for instance, Figure 12-13 of *Reservoir Stimulation, Third Edition*).

THE WELL-FRACTURE-RESERVOIR SYSTEM

We consider a fully penetrating vertical fracture in a pay layer of thickness h, as shown in Figure 3-1.

Note that in reality the drainage area is neither circular nor rectangular, however, for most drainage shapes these geometries are reasonable approximations. Using r_e or x_e is only a matter of convenience. The relation between the drainage area A, the drainage radius r_e and the drainage side length, x_e, is given by

$$A = r_e^2 \pi = x_e^2 \qquad (3\text{-}10)$$

For a vertical well intersecting a rectangular vertical fracture that penetrates fully from the bottom to the top of the rectangular drainage

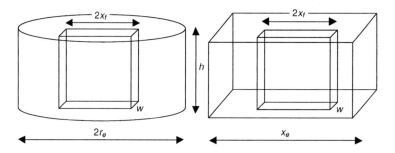

FIGURE 3-1. Notation for fracture performance.

volume, the performance is known to depend on the penetration ratio in the x direction,

$$I_x = \frac{2x_f}{x_e} \tag{3-11}$$

and on the dimensionless fracture conductivity,

$$C_{fD} = \frac{k_f w}{k x_f} \tag{3-12}$$

where x_f is the fracture half length, x_e is the side length of the square drainage area, k is the formation permeability, k_f is the proppant pack permeability, and w is the average (propped) fracture width.

PROPPANT NUMBER

The key to formulating a meaningful technical optimization problem is to realize that the fracture penetration and the dimensionless fracture conductivity (through width) are competing for the same resource: the propped volume. Once the reservoir and proppant properties and the amount of proppant are fixed, one has to make the optimal compromise between width and length. The available propped volume puts a constraint on the two dimensionless numbers. To handle the constraint easily, we introduce the dimensionless proppant number:

$$N_{prop} = I_x^2 C_{fD} \tag{3-13}$$

The proppant number as defined above is just a combination of the other two dimensionless parameters: penetration ratio and dimensionless fracture conductivity. Substituting the definition of the penetration ratio and dimensionless fracture conductivity into Equation 3-13, we obtain

$$N_{prop} = \frac{4k_f x_f w}{k x_e^2} = \frac{4k_f x_f wh}{k x_e^2 h} = \frac{2k_f}{k} \frac{V_{prop}}{V_{res}} \tag{3-14}$$

where N_{prop} is the proppant number, dimensionless; k_f is the effective proppant pack permeability, md; k is the formation permeability, md; V_{prop} is the propped volume in the pay (two wings, including void space between the proppant grains), ft^3; and V_{res} is the drainage volume (i.e., drainage area multiplied by pay thickness), ft^3. (Of course, any other coherent units can be used, because the proppant number involves only the ratio of permeabilities and the ratio of volumes.)

Equation 3-14 plainly reveals the meaning of the proppant number: it is the weighted ratio of propped fracture volume (two wings) to reservoir volume, with a weighting factor of two times the proppant-to-formation permeability contrast. Notice, only the proppant that reaches the pay layer is counted in the propped volume. If, for instance, the fracture height is three times the net pay thickness, then V_{prop} can be estimated as the bulk (packed) volume of injected proppant divided by three. In other words, the packed volume of the injected proppant multiplied by the volumetric proppant efficiency yields the V_{prop} used in calculating the proppant number.

The dimensionless proppant number, N_{prop}, is by far the most important parameter in unified fracture design.

Figure 3-2 shows J_D represented in a traditional manner, as a function of dimensionless fracture conductivity, C_{fD}, with I_x as a parameter. Similar graphs showing productivity increase are common in the published literature.

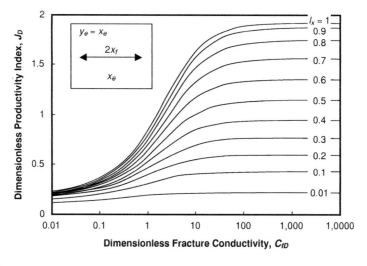

FIGURE 3-2. Dimensionless productivity index as a function of dimensionless fracture conductivity, with I_x as a parameter (McGuire-Sikora type representation).

However, Figure 3-2 is not very helpful in solving an optimization problem involving a fixed amount of proppant. For this purpose, in Figures 3-3 and 3-4, we present the same results, but now with the proppant number, N_{prop}, as a parameter. The individual curves correspond to J_D at a fixed value of the proppant number.

As seen from Figures 3-3 and 3-4, for a given value of N_{prop}, the maximum productivity index is achieved at a well-defined value of the dimensionless fracture conductivity. Because a given proppant number represents a fixed amount of proppant reaching the pay, the best compromise between length and width is achieved at the dimensionless fracture conductivity located under the peaks of the individual curves.

One of the main results seen from the figures is, that at proppant numbers less than 0.1, the optimal compromise occurs always at C_{fD} = 1.6. When the propped volume increases, the optimal compromise happens at larger dimensionless fracture conductivities, simply because the dimensionless penetration cannot exceed unity (i.e., once a fracture reaches the reservoir boundary, additional proppant is allocated only to fracture width). This effect is shown in Figure 3-4, as is the absolute maximum achievable dimensionless productivity index of 1.909. The maximum value of PI, equal to $6/\pi$, is the productivity index corresponding to perfect linear flow in a square reservoir.

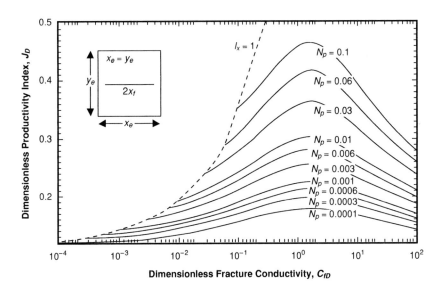

FIGURE 3-3. Dimensionless productivity index as a function of dimensionless fracture conductivity, with proppant number as a parameter (for $N_{prop} < 0.1$).

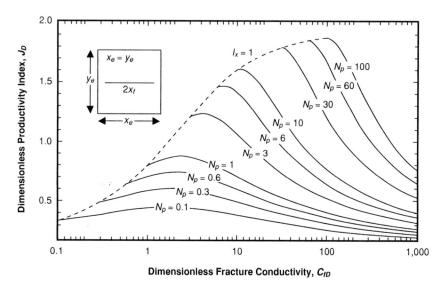

FIGURE 3-4. Dimensionless productivity index as a function of dimensionless fracture conductivity, with proppant number as a parameter (for $N_{prop} > 0.1$).

In medium and high permeability formations (above 50 md), it is practically impossible to achieve a proppant number larger than 0.1. For frac & pack treatments, typical proppant numbers range between 0.0001 and 0.01. Thus, for medium to high permeability formations, the optimum dimensionless fracture conductivity is always $C_{fDopt} = 1.6$.

In "tight gas" reservoirs, it is possible to achieve large dimensionless proppant numbers, at least in principle. Proppant numbers calculated for a limited drainage area—and not questioning the portion of proppant actually contained in the pay layer—can be as high as 1 to 10. However, in practice, proppant numbers larger than 1 may be difficult to achieve. For large treatments, the proppant can migrate upward, creating excessive and unplanned fracture height, or it might penetrate laterally outside of the assigned drainage area.

The situation is more complex for an individual well in a larger area. In this case, a (hypothetical) large fracture length tends to increase the drained reservoir volume, and the proppant number decreases. Ultimately, the large fracture is beneficial, but the achievable proppant number remains limited.

In reality, even trying to achieve proppant numbers larger than unity would be extremely difficult. Indeed, for a large proppant number, the optimum C_{fD} determines an optimum penetration ratio near unity. This can be easily seen from Figure 3-5, where the penetration

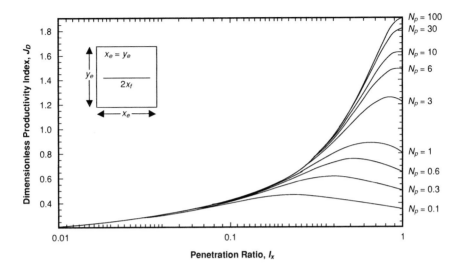

FIGURE 3-5. Dimensionless productivity index as a function of penetration ratio, with proppant number as a parameter (for $N_{prop} > 0.1$).

ratio is shown on the x-axis. To place the proppant "wall-to-wall" while keeping it inside the drainage volume would require a precision in the fracturing operation that is practically impossible to achieve.

The maximum possible dimensionless productivity index for $N_{prop} = 1$ is about $J_D = 0.9$. The dimensionless productivity index of an undamaged vertical well is between 0.12 and 0.14, depending on the well spacing and assumed well radius. Hence, there is a realistic maximum for the "folds of increase" in the pseudo-steady state productivity index (with respect to the zero skin case), i.e., 0.9 divided by 0.13 is approximately equal to 7. Larger folds of increase are not likely. Of course, larger folds of increase can be achieved with respect to an originally damaged well where the pre-treatment skin factor has a large and positive value.

Another common misunderstanding is related to the transient flow period. Under transient flow, the productivity index (and hence the production rate) is larger than in the pseudo-steady state case. With this qualitative picture in mind, it is easy to discard the pseudo-steady state optimization procedure and to "shoot for" very high dimensionless fracture conductivities and/or to anticipate many more folds of increase in the productivity. In reality, the existence of a transient flow period does not change the previous conclusions on optimal dimensions. Our calculations show that there is no reason to depart from the optimum compromise derived for the pseudo-steady state case, even if the well will produce in the transient regime for a considerable time (say months or even years). Simply stated, what is good for maximizing pseudo-steady state flow is also good for maximizing transient flow.

In the definition of proppant number, k_f is the effective (or equivalent, as it is sometimes called) permeability of the proppant pack. This parameter is crucial in design. Current fracture simulators generally provide a nominal value for the proppant pack permeability (supplied by the proppant manufacturer) and allow it to be reduced by a factor that the user selects. The already-reduced value should be used in the proppant number calculation.

There are numerous reasons why the actual (or equivalent) proppant pack permeability will be lower than the nominal value. The main reasons are as follows:

- Large closure stresses crush the proppant, reducing the average grain size, grain uniformity, and porosity.

- Fracturing fluid residue decreases the permeability in the fracture.

- High fluid velocity in the proppant pack creates "non-Darcy effects," resulting in additional pressure loss. This phenomenon can be significant when gas is produced in the presence of a liquid (water and/or condensate). The non-Darcy effect is caused by the periodic acceleration-deceleration of the liquid droplets, effectively reducing the permeability of the proppant pack. This reduced permeability can be an order of magnitude lower than the nominal value presented by the manufacturer.

During the fracture design, considerable attention must be paid to the effective permeability of the proppant pack and to the permeability of the formation. Knowledge of the effective permeability contrast is crucial, and cannot be substituted by qualitative reasoning.

Well Performance for Low and Moderate Proppant Numbers

By low and moderate proppant numbers, we mean anything less than 0.1. The most dynamic fracturing activities (frac & pack, for example) fall into this category—making it extremely important from a design standpoint.

The optimum treatment design for moderate proppant numbers can be simply and concisely presented in an analytical form. In the process, we will show how the proppant number and dimensionless productivity index relate to some other popular performance indicators, such as the Cinco-Ley and Samaniego *pseudo-skin function* and Prats' *equivalent wellbore radius*. In fact, fracture designs based on these related performance indicators are just the moderate (low) proppant number limit of the more comprehensive unified fracture design.

Prats (1961) introduced the concept of equivalent wellbore radius resulting from a fracture treatment. He also showed that, except for the fracture extent, all fracture variables affect well performance only through the combined quantity of dimensionless fracture conductivity. When the dimensionless fracture conductivity is high (e.g., greater than 100), the behavior is similar to that of an infinite conductivity fracture. The behavior of infinite conductivity fractures was studied later by Gringarten and Ramey (1974). To characterize the impact of a finite-conductivity vertical fracture on the performance of a vertical well, Cinco-Ley and Samaniego (1981) introduced a pseudo-skin function which is strictly a function of dimensionless fracture conductivity.

According to the definition of pseudo-skin factor, the dimensionless pseudo-steady state productivity index can be given as

$$J_D = \frac{1}{\ln 0.472 \dfrac{r_e}{r_w} + s_f} \tag{3-15}$$

where s_f is the pseudo-skin. In Prats' notation the same productivity index is described by

$$J_D = \frac{1}{\ln 0.472 \dfrac{r_e}{r_w'}} \tag{3-16}$$

where r_w' is the equivalent wellbore radius. Prats also used the *relative equivalent wellbore radius* defined by r_w' / x_f.

In the Cinco-Ley formalism, the productivity index is described as

$$J_D = \frac{1}{\ln 0.472 \dfrac{r_e}{x_f} + f} \tag{3-17}$$

where f is the pseudo-skin function with respect to the fracture half-length.

Table 3-2 shows the relations between these quantities.

The advantage of the Cinco-Ley and Samaniego formalism (f-factor) is that, for moderate (and low) proppant numbers, the quantity f depends only on the dimensionless fracture conductivity. The solid line in Figure 3-6 shows the Cinco-Ley and Samaniego f-factor as a function of dimensionless fracture conductivity.

Note that for large values of C_{fD}, the f-factor expression approaches ln(2), indicating that the production from an infinite conductivity fracture is equivalent to the production of $\pi/2$ times *more* than the production from the same surface arranged cylindrically (like the wall of a huge wellbore). In calculations, it is convenient to use an explicit expression of the form

$$f = \frac{1.65 - 0.328u + 0.116u^2}{1 + 0.18u + 0.064u^2 + 0.005u^3}, \quad \text{where} \quad u = \ln C_{fD} \qquad (3\text{-}18)$$

Because the relative wellbore radius of Prats can be also expressed by the f-factor (see Table 3-2), we obtain the equivalent result:

$$\frac{r'_w}{x_f} = \exp\left[-\frac{1.65 - 0.328u + 0.116u^2}{1 + 0.18u + 0.064u^2 + 0.005u^3}\right], \quad \text{where} \quad u = \ln C_{fD} \qquad (3\text{-}19)$$

The simple curve-fits represented by Equations 3-18 and 3-19 are only valid over the range indicated in Figure 3-6. For very large values of C_{fD}, one can simply use the limiting value for Equation 3-19, which is 0.5, showing that the infinite conductivity fracture has a productivity similar to an imaginary (huge) wellbore with radius $x_f/2$.

Interestingly enough, infinite conductivity behavior does not mean that we have selected the optimum way to place a given amount of proppant into the formation.

TABLE 3-2.   Relations Between Various Performance Indicators

$f = s_f + \ln\left[\dfrac{x_f}{r_w}\right]$	$s_f = \ln\left[\dfrac{r_w}{r'_w}\right]$
$r'_w = r_w \exp[-s_f]$	$r'_w = x_f \exp[-f]$
$\dfrac{r'_w}{x_f} = \exp[-f]$	$\dfrac{r'_w}{x_f} = \dfrac{r_w}{x_f}\exp[-s_f]$

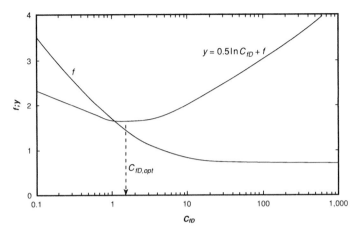

FIGURE 3-6. Cinco-Ley and Samaniego *f*-factor and the *y*-function.

OPTIMUM FRACTURE CONDUCTIVITY

In this context ($N_{prop} < 0.1$), a strictly physical optimization problem can be formulated: How should we select the length and width if the propped volume of one fracture wing, $V_f = w \times h \times x_f$, is given as a constraint, and we wish to maximize the PI in the pseudo-steady state flow regime. It is assumed that the formation thickness, drainage radius, and formation and proppant pack permeabilities are known, and that the fracture is vertically fully penetrating (i.e., $h_f = h$).

Selecting C_{fD} as the decision variable, the length is expressed as

$$x_f = \left(\frac{V_f k_f}{C_{fD} hk} \right)^{1/2} \tag{3-21}$$

Substituting Equation 3-21 into 3-17, the dimensionless productivity index becomes

$$J_D = \frac{1}{\ln 0.472 r_e + 0.5 \ln \dfrac{hk}{V_f k_f} + \left(0.5 \ln C_{fD} + f \right)} \tag{3-22}$$

where the only unknown variable is C_{fD}. Because the drainage radius, formation thickness, the two permeabilities, and the propped volume are fixed, the maximum PI occurs when the quantity in parentheses,

$$y = 0.5 \ln C_{fD} + f \tag{3-23}$$

reaches a minimum. That quantity is also shown in Figure 3-6. Because the above expression depends only on C_{fD}, the optimum, $C_{fD,opt} = 1.6$ is a given constant for any reservoir, well, and proppant volume.

This result provides a deeper insight to the real meaning of dimensionless fracture conductivity. The reservoir and the fracture can be considered as a system working in series. The reservoir can feed more fluids into the fracture if the length is larger, but (since the volume is fixed) this means a narrower fracture. In a narrow fracture, the resistance to flow may be significant. The optimum dimensionless fracture conductivity corresponds to the best compromise between the requirements of the two subsystems. Once it is found, the optimum fracture half-length can be calculated from the definition of C_{fD} as

$$x_f = \left(\frac{V_f k_f}{1.6 h k} \right)^{1/2} \tag{3-24}$$

and consequently, the optimum propped average width should be

$$w = \left(\frac{1.6 V_f k}{h k_f} \right)^{1/2} = \frac{V_f}{h x_f} \tag{3-25}$$

Notice that V_f is $V_{prop}/2$ because it is only one half of the propped volume.

The most important implication of the above results is that *there is no theoretical difference between low and high permeability fracturing.* In all cases, there exists a physically optimal fracture that should have a C_{fD} near unity. In low permeability formations, this requirement results in a long and narrow fracture; in high permeability formations, a short and wide fracture provides the same dimensionless conductivity.

If the fracture length and width are selected according to the optimum compromise, the dimensionless productivity index will be

$$J_{D,max} = \frac{1}{0.99 - 0.5 \ln N_{prop}} \tag{3-26}$$

Of course, the indicated optimal fracture dimensions may not be technically or economically feasible. In low permeability formations, the indicated fracture length may be too large, or the extreme narrow width may mean that the assumed constant proppant permeability is no longer valid. In high permeability formations, the indicated large width might be impossible to create. For more detailed calculations, all the constraints must be taken into account, but, in any case, a dimensionless fracture conductivity far from the optimum indicates that either the fracture is a relative "bottleneck" ($C_{fD} \ll 1.6$) or that it is too "short and wide" ($C_{fD} \gg 1.6$).

The reader should not forget that the results of this section— including the Cinco-Ley and Samaniego graph and its curve fit, the optimum dimensionless fracture conductivity of 1.6, and Equation 3-26— are valid only for proppant numbers less than 0.1. This can be easily seen by comparing Figures 3-3 and 3-4. In Figure 3-3, the curves have their maximum at $C_{fD} = 1.6$, and the maximum J_D corresponds to the simple Equation 3-26. In Figure 3-4, however, where the proppant numbers are larger than 0.1, the location of the maximum is shifted, and the simple calculations based on the f-factor (Equation 3-18) or on the equivalent wellbore radius (Equation 3-19) are no longer valid.

Optimization routines found on the CD that accompanies this book are based on the full information contained in Figures 3-3 and 3-4, and formulas developed for moderate proppant numbers are used only in the range of their validity.

DESIGN LOGIC

We wish to place a certain amount of proppant in the pay interval and to place it in such a way that the maximum possible productivity index is realized. The key to finding the right balance between size and productivity improvement is in the proppant number. Since V_{prop} includes only that part of the proppant that reaches the pay, and hence is dependent on the volumetric proppant efficiency, the proppant number cannot be simply fixed during the design procedure.

In unified fracture design, we specify the amount of proppant indicated for injection and then proceed as follows:

1. Assume a volumetric proppant efficiency and determine the proppant number. (Once the treatment details are obtained, the assumed volumetric proppant efficiency related to created fracture height may be revisited and the design process may be repeated in an iterative manner.)

2. Use Figure 3-3 or Figure 3-4 (or rather the design spreadsheet) to calculate the maximum possible productivity index, J_{Dmax}, and also the optimum dimensionless fracture conductivity, C_{fDopt}, from the proppant number.

3. Calculate the optimum fracture half-length. Denoting the volume of one propped wing (in the pay) by V_f, the optimum fracture half-length can be calculated as

$$x_f = \left(\frac{V_f k_f}{C_{fD,opt} hk} \right)^{1/2}$$

(3-27)

4. Calculate the optimum averaged propped fracture width as

$$w = \left(\frac{C_{fD,opt} V_f k}{hk_f} \right)^{1/2} = \frac{V_f}{x_f h}$$

(3-28)

In the above two equations, V_f and h must correspond to each other. If total fracture height is used for h, which is often denoted by h_f, then the proppant volume V_f must be the total propped volume of one wing. However, if the selected V_f corresponds only to that portion of one wing volume that is contained in the pay layer, then h should be the net thickness of the pay. The final result for optimum length and width will be the same in either case. It is a better practice, however, to use net thickness and net volume (contained in the pay) because those variables are also used to calculate the proppant number.

Once reservoir engineering and economic considerations have dictated the fracture dimensions to be created, the next issue is how to achieve that goal. From this point, design of the fracture treatment can be viewed as adjusting treatment details (pumping time and proppant schedule) to achieve the desired final fracture dimensions.

In the next chapter, we outline the mechanics of fracture creation in some detail. This theoretical basis is needed before we can proceed to design the fracture treatment, our ultimate goal.

Fracturing Theory

In the following, we briefly summarize the most important mechanical concepts related to hydraulic fracturing.

LINEAR ELASTICITY AND FRACTURE MECHANICS

Elasticity implies reversible changes. The initiation and propagation of a fracture means that the material has responded in an inherently non-elastic way, and an *irreversible* change has occurred. Nevertheless, linear elasticity is a useful tool when studying fractures because both the stresses and strains (except perhaps in the vicinity of the fracture face, and especially the tip) may still be adequately described by elasticity theory.

A linear elastic material is characterized by elastic constants that can be determined in static or dynamic loading experiments. For an isotropic material, where the properties are independent of direction, two constants are sufficient to describe the behavior.

Figure 4-1 is a schematic representation of a static experiment with uniaxial loading. The two parameters obtained from such an experiment are the Young's modulus (E) and the Poisson ratio (ν). They are

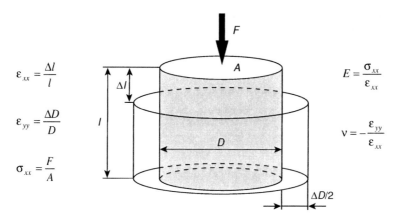

$$\varepsilon_{xx} = \frac{\Delta l}{l}$$

$$\varepsilon_{yy} = \frac{\Delta D}{D}$$

$$\sigma_{xx} = \frac{F}{A}$$

$$E = \frac{\sigma_{xx}}{\varepsilon_{xx}}$$

$$v = -\frac{\varepsilon_{yy}}{\varepsilon_{xx}}$$

FIGURE 4-1. Uniaxial loading experiment.

calculated from the vertical stress (σ_{xx}) vertical strain (ε_{xx}) and horizontal strain (ε_{yy}), as shown in the figure.

Table 4-1 shows the interrelation of those constants most often used in hydraulic fracturing. The plane strain modulus (E') is the only elastic constant really needed in our equations.

In linear elastic theory, the concept of *plane strain* is often used to reduce the dimensionality of a problem. It is assumed that the body is infinite in at least one direction, and external forces (if any) are applied parallel to this direction (i.e., "infinitely repeated" in every cross section). In such case, it is intuitively obvious that the state of strain also repeats itself infinitely.

TABLE 4-1. Interrelation of Various Properties of a Linear Elastic Material

Required/Known	E, v	G, v	E, G
Shear modulus, G	$\dfrac{E}{2(1+v)}$	G	G
Young's modulus, E	E	$2G(1+v)$	E
Poisson ratio, v	v	v	$\dfrac{E-2G}{2G}$
Plane strain modulus, E'	$\dfrac{E}{1-v^2}$	$\dfrac{2G}{1-v}$	$\dfrac{4G^2}{4G-E}$

Plane strain is a reasonable approximation in a simplified description of hydraulic fracturing. The main question is how to select the plane. Two possibilities arise, and, in turn, this has given rise to two different approaches to fracture modeling. The state of plane strain was assumed in the horizontal plane by Khristianovitch and Zheltov (1955) and by Geertsma and de Klerk (1969), while plane strain in the vertical plane (normal to the direction of fracture propagation) was assumed by Perkins and Kern (1961) and Nordgren (1972).

Often, in the hydraulic fracturing literature, the term "KGD" geometry is used interchangeably to the horizontal plane-strain assumption and "PKN" geometry is used as a substitute for postulating plane strain in the vertical plane.

Exact mathematical solutions are available for the problem of a pressurized crack in the state of plane strain. In particular, it is well known that the pressurized line crack has an elliptical width distribution (Sneddon, 1973):

$$w(x) = \frac{4p_0}{E'} \sqrt{c^2 - x^2} \tag{4-1}$$

where x is the distance from the center of the crack, c is the half-length (the distance of the tip from the center) and p_0 is the constant pressure exerted on the crack faces from inside. From Equation 4-1, the maximum width at the center is

$$w_0 = \frac{4cp_0}{E'} \tag{4-2}$$

indicating that a linear relationship is maintained between the crack opening induced and the pressure exerted. When the concept of pressurized line crack is applied for a real situation, p_0 is substituted with the net pressure, p_n, defined as the difference of the inner pressure and the minimum principal stress acting from outside, trying to close the fracture (Hubbert and Willis, 1957; Haimson and Fairhurst, 1967).

Fracture mechanics has emerged from the observation that any existing discontinuity in a solid deteriorates its ability to carry loads. A (possibly small) hole may give rise to high local stresses compared to the ones being present without the hole. The high stresses, even if they are limited to a small area, may lead to the rupture of the material. It is often convenient to look at material discontinuities as stress concentrators which locally increase the otherwise present stresses.

Two main cases must be distinguished. If the form of discontinuity is smooth (e.g., a circular borehole in a formation), then the maximum stress around the discontinuity is higher than the virgin stress by a finite factor, which depends on the geometry. For example, the stress concentration factor for a circular borehole is three.

The situation is different in the case of sharp edges, such as at the tip of a fracture. Then the maximum stress at the tip becomes infinite. In fracture mechanics, we have to deal with singularities. Two different loadings (pressure distributions) of a line crack result in two different stress distributions. Both cases may yield infinite stresses at the tip, but the "level of infinity" is different. We need a quantity to characterize this difference. Fortunately, all stress distributions near the tip of any fracture are similar in the sense that they decrease according to $r^{-1/2}$, where r is the distance from the tip. The quantity used to characterize the "level of infinity" is the stress intensity factor, K_I, defined as the multiplier to the $r^{-1/2}$ function. For the idealization of a pressurized line crack with half-length, c, and constant pressure, p_0, the stress intensity factor is given by

$$K_I = p_0 c^{1/2} \tag{4-3}$$

In other words, the stress intensity factor at the tip is proportional to the constant pressure opening up the crack and to the square root of the crack half-length (characteristic dimension).

According to the key postulate of linear elastic fracture mechanics (LEFM), for a given material there is a critical value of the stress intensity factor, K_{IC}, called fracture toughness. If the stress intensity factor at the crack tip is above the critical value, the crack will propagate; otherwise it will not. Fracture toughness is a useful quantity for safety calculations, when the engineer's only concern is to avoid fracturing. In well stimulation, where the engineer's primary goal is to create and propagate a fracture, the concept has been found somewhat controversial because it predicts that less and less effort is necessary to propagate a fracture with increasing extent. In the large scale, however, the opposite is usually true.

FRACTURING FLUID MECHANICS

Fluid materials deform continuously (in other words, flow) without rupture when subjected to a constant stress. Solids generally will assume a static equilibrium deformation under the same stresses. Crosslinked fracturing fluids usually behave as viscoelastic fluids. Their stress-strain material functions fall between those of pure fluids and solids.

From our point of view, the most important property of fluids is their resistance to flow. The local intensity of flow is characterized by the shear rate, $\dot{\gamma}$, measured in 1/s. It can be considered as the rate of change of velocity with the distance between sliding layers. The stress emerging between the layers is the shear stress, τ. Its dimension is force per unit area (in SI units, Pa). The material function relating shear stress and shear rate is the *rheological curve*. This information is necessary to calculate the pressure drop (actually, energy dissipation) for a given flow situation, such as flow in pipe or flow between parallel plates.

Apparent viscosity is defined as the ratio of stress to shear rate. Generally, the apparent viscosity varies with shear rate, except in the case of a Newtonian fluid—a very specific fluid in which the viscosity is a constant. The rheological curve and the apparent viscosity curve contain the same information and are used interchangeably. Figure 4-2 shows typical rheological curves, and Table 4-2 lists some commonly used rheological constitutive equations.

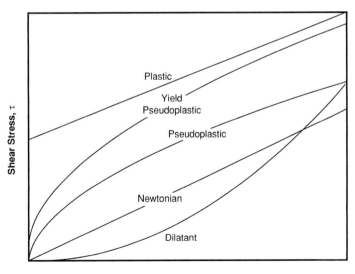

FIGURE 4-2. Typical rheological curves.

TABLE 4-2.   **Commonly Used Rheological Constitutive Equations**

$\tau = \mu \dot\gamma$	Newtonian
$\tau = K \dot\gamma^n$	Power law
$\tau = \tau_y + \mu_p \dot\gamma$	Bingham plastic
$\tau = \tau_y + K \dot\gamma^n$	Yield power law

The model parameters vary with chemical composition, temperature and, to a lesser extent, many other factors including shear history. In the case of foams, the volumetric ratio between the gas and liquid phases plays an important role (Reidenbach, 1985; Winkler, 1995).

Most fracturing gels exhibit significant shear thinning (i.e., loss of viscosity with increasing shear rate). A constitutive equation that captures this primary aspect of their flow behavior is the Power law model. The flow behavior index, n, usually ranges from 0.3 to 0.6.

All fluids exhibit some finite limiting viscosity at high shear rates. The build-up of very high apparent viscosity at low shear might be approximated by the inclusion of a yield stress for certain fluids. Many fluids demonstrate what appears to be Newtonian behavior at low shear rates.

Much of the current rheology research focuses on building more realistic apparent viscosity models that effectively incorporate each of the previously mentioned characteristics as well as the nonlinear, time-dependent viscoelastic effects of crosslinked gels.

A rheological model is used to predict the pressure losses (gradient) associated with an average fluid flow velocity in a given physical geometry. The equations of motion have been solved for the standard rheological models in the most obvious geometries (e.g., flow in circular tubes, annuli, and between thin-gap parallel plates). The solution is often presented as a relation between average linear velocity (flow rate per unit area) and pressure drop. In calculations, it is convenient to use the equivalent Newtonian viscosity (μ_e), that is, the viscosity that would be used in the equation of the Newtonian fluid to obtain the same pressure drop under the same flow conditions. While apparent viscosity (at a given local shear rate) is the property of the fluid, equivalent viscosity depends also on the flow geometry and carries the same information as the pressure drop solution. For more complex rheological models, there is no closed-form solution (neither for the pressure drop nor for the equivalent Newtonian viscosity), and the calculations involve numerical root-finding.

Of particular interest to hydraulic fracturing is the laminar flow in two limiting geometries. *Slot flow* occurs in a channel of rectangular cross section when the ratio of the longer side to the shorter side is extremely large. Limiting *ellipsoid flow* occurs in an elliptic cross section with extremely large aspect ratio. The former corresponds to the KGD geometry and the latter to the PKN geometry.

Table 4-3 gives the solutions commonly used in hydraulic fracturing calculations. The most familiar equation, valid for Newtonian behavior, is presented first. Then an equivalent viscosity is given for the Power law fluid. The equivalent viscosity can be used in the Newtonian form of the pressure drop equation. Notice that the equivalent viscosity depends on the average velocity (u_{avg}) and on the geometry of the flow channel (in case of slot flow, on the width, w; in case of elliptical cross section, on the maximum width, w_0). It is interesting to note that the equation for laminar flow of a Power law fluid in the limiting ellipsoid geometry has not been derived. The solution presented here can be obtained by analogy considerations (for details, see Valkó and Economides, 1995).

The friction pressure associated with pumping fracturing fluids through surface lines and tubulars cannot be calculated directly using the classic turbulent flow correlations. Special relations have to be applied to account for the drag reduction phenomena caused by the long polymer chains. Rheological behavior also plays an important role in the proppant carrying capacity of the fluid (Roodhart, 1985; Acharya, 1986).

TABLE 4-3.   Pressure Drop and Equivalent Newtonian Viscosity

Rheological model	Newtonian $\tau = \mu\dot{\gamma}$	Power law $\tau = K\dot{\gamma}^n$
Slot flow	$\dfrac{\Delta p}{L} = \dfrac{12\mu u_{avg}}{w^2}$	$\mu_e = \dfrac{2^{n-1}}{3}\left(\dfrac{1+2n}{n}\right)^n Kw^{1-n}u_{avg}^{n-1}$
Ellipsoid flow	$\dfrac{\Delta p}{L} = \dfrac{16\mu u_{avg}}{w_0^2}$	$\mu_e = \dfrac{2^{n-1}}{\pi}\left[\dfrac{1+(\pi-1)n}{n}\right]^n Kw_0^{1-n}u_{avg}^{n-1}$

LEAKOFF AND VOLUME BALANCE IN THE FRACTURE

The polymer content of the fracturing fluid is partly intended to impede the loss of fluid into the reservoir. This phenomenon is envisaged as a continuous build-up of a thin layer of polymer (the filter cake), which manifests an ever-increasing resistance to flow through the fracture face. The actual leakoff is determined by a coupled system that includes not only the filter cake, which is one element, but also flow conditions in the reservoir.

A fruitful approximation dating back to Carter, 1957 (cf. appendix to Howard and Fast, 1957), is to consider the combined effect of the different phenomena as a material property. According to this concept, the leakoff velocity, v_L, is given by the Carter I equation:

$$v_L = \frac{C_L}{\sqrt{t}} \tag{4-4}$$

where C_L is the leakoff coefficient (length/time$^{1/2}$) and t is the time elapsed since the start of the leakoff process. The integrated form of the Carter equation is

$$\frac{V_{Lost}}{A_L} = 2C_L\sqrt{t} + S_p \tag{4-5}$$

where V_{Lost} is the fluid volume that passes through the surface A_L during the time period from time zero to time t. The integration constant, S_P, is called the *spurt loss coefficient*. It can be considered as the width of the fluid body passing through the surface instantaneously at the very beginning of the leakoff process. Correspondingly, the term $2C_L\sqrt{t}$ can be considered as the leakoff width. (Note that the factor 2 is an artifact of the integration. It has nothing to do with the "two wings" and/or "two faces" introduced later.) The two coefficients, C_L and S_P, can be determined from laboratory tests or, preferably, from evaluation of a fracture calibration test.

Formal Material Balance: The Opening-Time Distribution Factor

Consider the fracturing treatment shown schematically in Figure 4-3. The volume V_i injected into one wing during the injection time t_e consists of two parts: the volume of one fracture wing at the end of

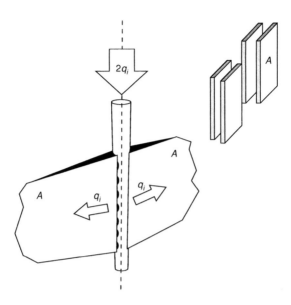

FIGURE 4-3. Notation for material balance.

pumping (V_e) and the volume lost (leakoff volume). The subscript e denotes that a given quantity is being measured or referenced at the end of pumping. Note that all the variables are defined with respect to one wing. The area A_e denotes the surface of one face of one fracture wing. Fluid efficiency η_e is defined as the fraction of the fluid remaining in the fracture: $\eta_e = V_e/V_i$. The average width, $\overline{w}$, is defined by the relation, $V = A\overline{w}$.

A hydraulic fracturing operation may last from tens-of-minutes up to several hours. Points on the fracture face near the well are "opened" at the beginning of pumping while the points near the fracture tip are younger. Application of Equation 4-5 necessitates the tracking of the opening-time of the different fracture face elements.

If only the overall material balance is considered, it is natural to rewrite the injected volume as the sum of the fracture volume, leakoff volume, and spurt volume using the formalism,

$$V_i = V_e + K_L \left(2A_e\, C_L \sqrt{t_e} \right) + 2A_e S_p \tag{4-6}$$

where the variable K_L is the opening-time distribution factor. It reflects the history of the evolution of the fracture surface, or rather the distribution of the opening-time, hence the name. In particular, if all the surface is opened at the beginning of the injection, then K_L

reaches its absolute maximum, $K_L = 2$. The fluid efficiency is the ratio of the created volume to the injected volume. Dividing both volumes by the final fracture area, we can consider fracture efficiency as the ratio of the created width to the would-be width, where the would-be width is defined as the sum of the created and lost widths.

Therefore, another form of Equation 4-6 is

$$\eta_e = \frac{\overline{w}_e}{\overline{w}_e + 2K_L C_L \sqrt{t_e} + 2S_p} \tag{4-7}$$

showing that the term $2K_L C_L \sqrt{t_e}$ can be considered as the "leakoff width," and the term $2S_p$ as the "spurt width." Equation 4-7 can be rearranged to obtain the opening-time distribution factor in terms of fluid efficiency and average width at the end of pumping:

$$K_L = -\frac{S_p}{C_L \sqrt{t_e}} - \frac{\overline{w}_e}{2C_L \sqrt{t_e}} + \frac{\overline{w}_e}{2\eta_e C_L \sqrt{t_e}} \tag{4-8}$$

Note that these relations are independent of the actual shape of the fracture face or the history of its evolution.

Constant Width Approximation (Carter Equation II)

In order to obtain an analytical solution for constant injection rate, Carter considered a hypothetical case in which the fracture width remains constant during the fracture propagation (the width "jumps" to its final value in the first instant of pumping). Then a closed form expression can be given for the fluid efficiency in terms of the two leakoff parameters and the width:

$$\eta_e = \frac{\overline{w}_e(\overline{w}_e + 2S_p)}{4\pi C_L^2 t_e}\left[\exp(\beta^2)\operatorname{erfc}(\beta) + \frac{2\beta}{\sqrt{\pi}} - 1\right] \tag{4-9}$$

where $\beta = \dfrac{2C_L\sqrt{\pi t_e}}{\overline{w}_e + 2S_p}$.

Power Law Approximation to Surface Growth

A basic assumption postulated by Nolte (1979, 1986) leads to a remarkably simple form of the material balance. He assumed that the fracture surface evolves according to a power law,

$$A_D = t_D^\alpha \tag{4-10}$$

where $A_D = A/A_e$ and $t_D = t/t_e$, and the exponent α remains constant during the entire injection period. Nolte realized that, in this case, the opening-time distribution factor is a function of α only. He represented the opening-time distribution factor and its dependence on the exponent of fracture surface growth using the notation $g_0(\alpha)$ and presented g_0 for selected values of α. A simple expression first obtained by Hagel and Meyer (1989) can be used to obtain the value of the opening-time distribution factor for any α:

$$g_0(\alpha) = \frac{\sqrt{\pi}\alpha\Gamma(\alpha)}{\Gamma(\alpha + 2/3)} \tag{4-11}$$

where $\Gamma(\alpha)$ is the Euler gamma function.

In calculations, the following approximation to the g_0 function might be easier to use:

$$g_0(\alpha) = \frac{2 + 2.06798\,\alpha + 0.541262\,\alpha^2 + 0.0301598\,\alpha^3}{1 + 1.6477\,\alpha + 0.738452\,\alpha^2 + 0.0919097\,\alpha^3 + 0.00149497\,\alpha^4} \tag{4-12}$$

Nolte assumed that the exponent remains between 0.5 and 1. With this assumption, the factor K_L lies between 4/3 (1.33) and $\pi/2$ (1.57), indicating that for two extremely different surface growth histories, the opening-time distribution factor varies less than 20 percent. Generally, the simple approximation $K_L = 1.5$ should provide enough accuracy for design purposes.

Various practitioners have related the exponent α to fracture geometry, fluid efficiency at the end of pumping, and fluid rheological behavior. None of these relations can be considered as proven theoretically, but they are reasonable engineering approximations, especially because the effect of the exponent on the final results is limited. Our recommendation is to use $\alpha = 4/5$ for the PKN, $\alpha = 2/3$ for the KGD, and $\alpha = 8/9$ for the radial model. These exponents can be derived from the no-leakoff equations shown later in Table 4-4.

Numerically, the original constant-width approximation of Carter and the power law surface growth assumption of Nolte give very similar results when used for design purposes. The g_0-function approach does, however, have technical advantages when applied to the analysis of calibration treatments.

Detailed Leakoff Models

The bulk leakoff model is not the only possible interpretation of the leakoff process. Several mechanistic models have been suggested in the past (Williams, 1970 and Settari, 1985; Ehlig-Economides, et al., 1994; Yi and Peden, 1994; Mayerhofer, et al., 1995). The total pressure difference between the inside of a created fracture and a far point in the reservoir is written as the sum,

$$\Delta p(t) = \Delta p_{face}(t) + \Delta p_{piz}(t) + \Delta p_{res}(t) \tag{4-13}$$

where Δp_{face} is the pressure drop across the fracture face dominated by the filter cake, Δp_{piz} is the pressure drop across a polymer-invaded zone and Δp_{res} is the pressure drop in the reservoir. Depending on their significance under the given conditions, one or two terms may be neglected. While the first two terms are connected to the leakoff rate at a given time instant, the reservoir pressure drop is transient. It depends on the entire history of the leakoff process, not only on its instant intensity.

The detailed leakoff models hold an advantage in that they are based on physically meaningful parameters, such as permeability and filter cake resistance, and they allow for explicit pressure-dependent simulation of the leakoff process. However, the application of these models is limited by the complexity of the mathematics involved and by the extra input they require.

BASIC FRACTURE GEOMETRIES

Engineering models for the propagation of a hydraulically induced fracture combine elasticity, fluid flow, material balance, and (in some cases) an additional propagation criterion. Given the fluid injection history, a model should predict the evolution with time of the fracture dimensions and the wellbore pressure.

For design purposes, an approximate description of the geometry might be sufficient, so simple models that predict fracture length and average width at the end of pumping are very useful. Models that predict these two dimensions—while the third one, fracture height, is fixed—are referred to as 2D models. If the fracture surface is postulated to propagate in a radial fashion, that is, the height is not fixed, the model is still considered to be 2D (the two dimensions being fracture radius and width).

A further simplification occurs if we can relate fracture length and width, neglecting the details of leakoff for now. This is the basic concept of the early so-called "width equations." It is assumed that the fracture evolves in two identical wings, perpendicular to the minimum principal stress of the formation. Because the minimum principal stress is usually horizontal (except for very shallow formations), the fracture will be vertical.

Perkins-Kern Width Equation

The PKN model assumes that the condition of plane strain holds in every vertical plane normal to the direction of propagation; however, unlike the rigorous plane-strain situation, the stress and strain state are not exactly the same in subsequent planes. In other words, the model applies a quasi-plane-strain assumption, and the reference plane is vertical, normal to the propagation direction. Neglecting the variation of pressure along the vertical coordinate, the net pressure, p_n, is considered as a function of the lateral coordinate x. The vertically constant pressure at a given lateral location gives rise to an elliptical cross section. Straightforward application of Equation 4-1 provides the maximum width of the ellipse as

$$w_0 = \frac{2 h_f \, p_n}{E'} \tag{4-14}$$

Perkins and Kern (1961) postulated that the net pressure is zero at the tip of the fracture, and they approximated the average linear velocity of the fluid at any location based on the one-wing injection rate (q_i) divided by the cross-sectional area. They obtained the pressure loss equation in the form,

$$\frac{dp_n}{dx} = -\frac{4\mu q_i}{\pi w_0^3 h_f} \tag{4-15}$$

Combining Equations 4-14 and 4-15, and integrating with the zero net pressure condition at the tip, they obtained the width profile:

$$w_0(x) = w_{w,o} \left(1 - \frac{x}{x_f} \right)^{1/4} \tag{4-16}$$

where the maximum width of the ellipse at the wellbore (see Figure 4-4) is given by

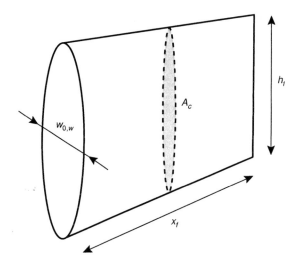

FIGURE 4-4. Basic notation for Perkins-Kern differential model.

$$w_{w,0} = 3.57 \left(\frac{\mu q_i x_f}{E'} \right)^{1/4} \tag{4-17}$$

In reality, the flow rate in the fracture is less than the injection rate, not only because part of the fluid leaks off, but also because the increase of width with time "consumes" another part of the injected fluid. In fact, what is more or less constant along the lateral coordinate at a given time instant, is not the flow rate, but rather the flow velocity, u_{avg}. However, repeating the Perkins-Kern derivation with a constant flow velocity assumption has very little effect on the final results.

Equation 4-17 is the Perkins-Kern width equation. It shows the effect of the injection rate, viscosity, and shear modulus on the width, once a given fracture length is achieved. Knowing the maximum width at the wellbore, we can calculate the average width, multiplying it by a constant shape factor, γ:

$$\overline{w} = \gamma w_{w,0}, \quad \text{where} \quad \gamma = \frac{\pi}{4} \frac{4}{5} = \frac{\pi}{5} = 0.628 \tag{4-18}$$

The shape factor contains two elements. The first one is $\pi/4$, which takes into account that the vertical shape is an ellipse. The second element is 4/5, which accounts for lateral variation in the maximum width.

In the petroleum industry, a version of Equation 4-17 with a slightly different constant is used more often, and is referred to as the Perkins-Kern-Nordgren (PKN) width equation (Nordgren, 1972):

$$w_{w,0} = 3.27 \left(\frac{\mu q_i x_f}{E'} \right)^{1/4} \tag{4-19}$$

Khristianovich-Zheltov-Geertsma-deKlerk Width Equation

The first model of hydraulic fracturing, elaborated by Khristianovich and Zheltov (1955), envisioned a fracture with the same width at any vertical coordinate within the fixed height, h_f. The underlying physical hypothesis is that the fracture faces slide freely at the top and bottom of the layer. The resulting fracture cross section is a rectangle. The width is considered as a function of the coordinate x. It is determined from the plane-strain assumption, now applied in the (every) horizontal plane. The Khristianovich and Zheltov model contained another interesting assumption: the existence of a non-wetted zone near the fracture tip. Geertsma and deKlerk (1969) accepted the main assumptions of Khristianovich and Zheltov and reduced the model into an explicit width formula. The KGD width equation is

$$w_w = \left(\frac{336}{\pi} \right)^{1/4} \left(\frac{\mu q_i x_f^2}{E' h_f} \right)^{1/4} = 3.22 \left(\frac{\mu q_i x_f^2}{E' h_f} \right)^{1/4} \tag{4-20}$$

In this case, the shape factor, relating the average width to the wellbore width, has no vertical component. Then, because of the elliptical horizontal shape, we obtain

$$\overline{w} = \gamma w_w, \quad \text{where} \quad \gamma = \frac{\pi}{4} = 0.785 \tag{4-21}$$

Daneshy's (1978) extension of the KGD model considers a non-constant pressure distribution along the fracture length, and a non-Newtonian fracturing fluid whose properties can change with time and temperature. Numerical computations yield the specific leakoff, increase in width, and flow rate at points along the fracture length during fracture extension.

For short fractures, where $2x_f < h_f$, the horizontal plane-strain assumption (KGD geometry) is more appropriate, and for $2x_f > h_f$, the vertical plane-strain assumption (PKN geometry) is physically more sound. Interestingly, for the special case when the total fracture length and height are equivalent, the two equations give basically the same average width and, hence, fracture volume.

Radial (Penny-shaped) Width Equation

This situation corresponds to horizontal fractures from vertical wells, vertical fractures extending from horizontal wells, or when fracturing relatively thick homogeneous formations—from a limited perforation interval in all cases. While the computations of fracture width are sensitive to how the fluid enters the fracture (a true point source would give rise to infinite pressure), a reasonable model can be postulated by analogy, which results in the same average width as the Perkins-Kern equation when $R_f = x_f = h_f/2$.

The result is

$$\overline{w} = 2.24\left(\frac{\mu q_i R_f}{E'}\right)^{1/4}$$

(4-22)

The real significance of the simple models presented in this section is the insight they provide—helping us to consider the effect of input data on the evolving fracture. Additional insight can be gained by comparing the fracture geometry and net pressure behavior of the models. Table 4-4 provides a direct side-by-side comparison of the basic fracture models (no-leakoff case).

The last row in Table 4-4 deserves particular attention. For the no-leakoff case, net pressure increases with time for the Perkins-Kern model, but decreases with time for the other two models. This is a well-known result that raises some questions. For example, in massive hydraulic fracturing, the net treating pressure most often increases with time, so net pressures derived from the Geertsma-deKlerk and radial models are of limited practical value. A more startling (and less well-known) observation is that the net pressures provided by the Geertsma-deKlerk and radial models are independent of injection rate. The KGD (and radial) view implies that when the fracture extent becomes large, very low net pressures are required to maintain a certain width. While this is a consequence of linear elasticity theory and the way that the plane-strain assumption is applied, it leads to absurd

TABLE 4-4. No-Leakoff Solutions of the Basic Fracture Models

Model	Perkins and Kern	Geertsma and deKlerk	Radial
Fracture Extent	$x_f = c_1 t^{4/5}$	$x_f = c_1 t^{2/3}$	$R_f = c_1 t^{4/9}$
	$c_1 = c_1' \left(\dfrac{q_i^3 E'}{\mu h_f^4} \right)^{1/5}$	$c_1 = c_1' \left(\dfrac{q_i^3 E'}{\mu h_f^3} \right)^{1/6}$	$c_1 = c_1' \left(\dfrac{q_i^3 E'}{\mu} \right)^{1/9}$
	$c_1' = \left(\dfrac{625}{512\pi^3} \right)^{1/5} = 0.524$	$c_1' = \left(\dfrac{16}{21\pi^3} \right)^{1/6} = 0.539$	$c_1' = 0.572$
Width	$w_{w,0} = c_2 t^{1/5}$	$w_w = c_2 t^{1/3}$	$w_{w,0} = c_2 t^{1/9}$
	$c_2 = c_2' \left(\dfrac{q_i^2 \mu}{E' h_f} \right)^{1/5}$	$c_2 = c_2' \left(\dfrac{q_i^3 \mu}{E' h_f} \right)^{1/6}$	$c_2 = c_2' \left(\dfrac{q_i^3 \mu^2}{E'^2} \right)^{1/9}$
	$c_2' = \left(\dfrac{2560}{\pi^2} \right)^{1/5} = 3.04$	$c_2' = \left(\dfrac{5376}{\pi^3} \right)^{1/6} = 2.36$	$c_2' = 3.65$
	$\bar{w} = \gamma w_{w,0}$	$\bar{w} = \gamma w_w$	$\bar{w} = \gamma w_{w,0}$
	$\gamma = 0.628$	$\gamma = 0.785$	$\gamma = 0.533$
Net Pressure	$p_{n,w} = c_3 t^{1/5}$	$p_{n,w} = c_3 t^{-1/3}$	$p_{n,w} = c_3 t^{-1/3}$
	$c_3 = c_3' \left(\dfrac{E'^4 \mu q_i^2}{h_f^6} \right)^{1/5}$	$c_3 = c_3' \left(E'^2 \mu \right)^{1/3}$	$c_3 = c_3' \left(E'^2 \mu \right)^{1/3}$
	$c_3' = \left(\dfrac{80}{\pi^2} \right)^{1/4} = 1.52$	$c_3' = \left(\dfrac{21}{16} \right)^{1/3} = 1.09$	$c_3' = 2.51$

results in the large scale. It is safe to say that the PKN model captures the physical fracturing process better than the other two models.

While many investigations have been performed during the last half century, the same ingredients must always appear in the "mix" of any suggested fracture model: material balance, relating injection rate and fracture volume; linear elasticity, relating fracture width to fracture extent; and fluid mechanics, relating width and pressure loss along the fracture. Additionally, an explicit fracture propagation criterion may be or may not be present.

Fracturing of High Permeability Formations

THE EVOLUTION OF THE TECHNIQUE

As recently as the early 1990s, hydraulic fracturing was used almost exclusively for low permeability reservoirs. The large fluid leakoff and unconsolidated sands associated with high permeability formations would ostensibly prevent the initiation and extension of a single, planar fracture with sufficient width to accept a meaningful proppant volume. Moreover, such fracture morphology, even if successfully created and propped, would be incompatible with the defined needs of moderate to high permeability reservoirs, that is, large conductivity (width).

A key breakthrough tied to the advance of high permeability fracturing (HPF) is the tip screenout (TSO), which arrests lateral fracture growth and allows for subsequent fracture inflation and packing. The result is short but wide to exceptionally wide fractures. While in traditional, unrestricted fracture growth an average fracture width of 0.25 in. would be considered normal, in TSO treatments, widths of one inch or even larger are commonly expected.

The role of hydraulic fracturing has expanded to encompass oil wells with permeabilities greater than 50 md and gas wells with over 5 md of permeability (Table 5-1). These wells clearly require a TSO design. Because of these developments, hydraulic fracturing has

TABLE 5-1. Fracturing Role Expanded

Permeability	Gas	Oil
Low	$k < 0.5$ md	$k < 5$ md
Moderate	$0.5 < k < 5$ md	$5 < k < 50$ md
High	$k > 5$ md	$k > 50$ md

captured an enormous share of all well completions, and further gains are certain, only tempered by the economy of scale affecting many petroleum provinces. In places such as the United States and Canada, hydraulic fracturing is poised to be applied to almost all petroleum wells drilled, as was shown in Figure 1-2.

It is interesting that HPF, which is often referred to as *frac & pack* or *fracpac*, did not necessarily originate as an extension of hydraulic fracturing—although HPF borrowed heavily from established techniques—but rather as a means of sand production control.

In controlling the amount of sand production to the surface, there are two distinctly different activities that can be done downhole: *sand exclusion* and *sand deconsolidation control*. Sand exclusion refers to all filtering devices such as screens and gravel packs. Gravel packing, the historically preferred well completion method to remedy sand production, is one such technique. These techniques do not prevent sand migration in the reservoir, so fines migrate and lodge in the gravel pack and screen, causing large damage skin effects. Well performance progressively deteriorates and often is not reversible with matrix stimulation treatments. Attempts to stem the loss in well performance by increasing the pressure drawdown often aggravates the problem further and may potentially lead to wellbore collapse.

A more robust approach is the control of sand deconsolidation, (i.e., prevention of fines migration at the source). It is widely perceived that the use of HPF accomplishes this by mating with the formation in its (relative) undisturbed state and reducing fluid velocities or "flux" at the formation face.

There are actually three factors that contribute to sand deconsolidation: (1) pressure drawdown and the "flux" created by the resulting fluid production, (2) the strength of the rock and integrity of the natural cementation, and (3) the state of stress in the formation. Of these three, the only factor that can be readily altered is the distribution of flow and pressure drawdown. By introducing formation fluids to the well along a more elongated path (e.g., a hydraulic fracture or horizontal well), it is entirely possible to reduce the fluid flux and, in turn, control sand production.

Consider a simple example by assuming a well that penetrates a 100 ft thick reservoir. If the well has a diameter equal to 1 ft, then the area for incoming radial flow in an open hole completion would be about 300 ft^2. However, for a fracture half-length of 100 ft, the area of flow would be $(2 \times 100 \times 100 \times 2)$ 40,000 ft^2. (Note: the second 2 accounts for the two walls of the fracture.) Remember that in a fractured well almost all fluid flow would be from the reservoir into the fracture, and then along the fracture into the well. For the same production rate, this calculation suggests the fluid flux in a fractured well would be less than 1/100th the fluid flux in an unfractured well.

Of course, not a great deal can be done to affect the state of stress or formation competence. The magnitude of earth stresses depends primarily on reservoir depth and to some extent pressure, with the situation becoming more complicated at depths of 3,000 ft or less. Pressure maintenance with gas or water flooding may be counter-productive unless maintenance of reservoir pressure allows economic production at a smaller drawdown. Various innovations have been suggested to remedy incompetent formations or improve on natural cementation—for example, by introducing complex well configurations or various exotic chemical treatments—but there is little that can be done to control this factor either.

In light of the discussion above, it should not be surprising that HPF has replaced gravel packs in many petroleum provinces susceptible to sand production, especially in operations where more sophisticated engineering is done. As with any stimulation technique that results in a productivity index improvement (defined as the production rate divided by the pressure drawdown), it is up to the operator to allocate this new productivity index either to a *larger rate* or a *lower drawdown,* or any combination of the two.

HPF indicates a marked departure from the heritage of gravel packing, incorporating more and more from hydraulic fracture technology. This trend can be seen, for instance, in the fluids and proppants applied. While the original *fracpack* treatments involved sand sizes and "clean" fluids common to gravel packing, the typical proppant sizes for hydraulic fracturing (20/40 mesh) now dominate. The increased application of crosslinked fracturing fluids also illustrates the trend.

For this reason, the terminology of "high permeability fracturing," or HPF, seems more appropriate than *fracpack,* and is used throughout this book.

In the following section, HPF is considered in a semi-quantitative light in view of competing technologies. This is followed by a discussion of the key issues in high permeability fracturing, including design, execution, and evaluation.

HPF IN VIEW OF COMPETING TECHNOLOGIES

Gravel Pack

Gravel pack refers to the placement of gravel (actually, carefully selected and sized sand) between the formation and the well in order to filter out (retain) reservoir particles that migrate through the porous medium. A "screen" is employed to hold the gravel pack in place. This manner of excluding reservoir fines from flowing into the well invariably causes an accumulation of fines in the near-well zone and a subsequent reduction in the gravel pack permeability (i.e., damage is caused).

The progressive deterioration of gravel pack permeability (increased skin effect) leads, in turn, to a decline in well production. Increasing pressure drawdown to counteract production losses can result in accelerated pore-level deconsolidation and additional sand production.

Any productivity index relationship (e.g., the steady-state expression for oil) can be used to demonstrate this point:

$$J = \frac{q}{p_e - p_{wf}} = \frac{kh}{141.2B\mu\left(\ln\dfrac{0.472r_e}{r_w} + s\right)} \tag{5-1}$$

Assuming $k = 50$ md, $h = 100$ ft, $B = 1.1$ res bbl/STB, $\mu = 0.75$ cp and $\ln r_e/r_w = 8.5$, the productivity indexes for an ideal (undamaged), a relatively damaged (e.g., $s = 10$), and a typical gravel packed well (e.g., $s = 30$) would be 5, 2.3, and 1.1 STB/d/psi, respectively. For a drawdown of 1,000 psi, these productivity indexes would result in production rates of 5,000, 2,300, and 1,100 STB/d, respectively. Clearly, the difference in production rates between the ideal and gravel packed wells can be considerable and very undesirable.

Consider for a moment the use of high permeability fracturing under the same scenario. This technology combines the advantages of propped fracturing to bypass the near-wellbore damage and gravel packing to provide effective sand control. Figure 5-1 is the classic presentation (compare Figure 3-6) of the equivalent skin effect (Cinco and Samaniego, 1978) in terms of dimensionless fracture conductivity, C_{fD} (= $k_f w/k x_f$), and fracture half-length, x_f.

It can be seen from Figure 5-1 that even with a hydraulic fracture of less than optimum conductivity (e.g., C_{fD} = 0.5) and short fracture length (e.g., x_f = 50 ft), the skin effect, s_f (again using r_w = 0.328 ft), would be equal to –3.

A negative skin effect equal to –3 applied to Equation 5-1 yields a productivity index of 7.7 STB/d/psi, more than a 50 percent increase over the ideal PI and seven times the magnitude of a damaged gravel-packed well. Even with a damaged fracture (e.g., leakoff-induced damage as described by Mathur et al., 1995) and a skin equal to –1, the productivity index would be 5.6 STB/d/psi, a five-fold increase over a damaged gravel-packed well.

This calculation brings forward a simple, yet frequently overlooked, issue. Small negative skin values have a much greater impact on well performance than comparable magnitudes (absolute value) of

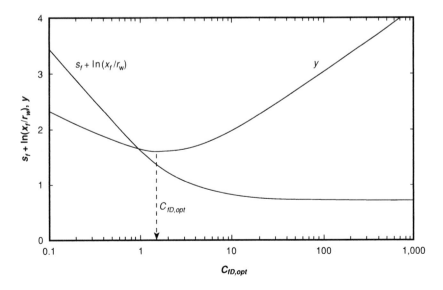

FIGURE 5-1. Pseudoskin factor for a vertical well intersected by a finite conductivity fracture.

positive skin. Furthermore, in the example calculation here, a five-fold increase in the productivity index suggests that the production rate would increase by the same amount if the drawdown is held constant. Under an equally possible scenario, the production rate could be held constant and the drawdown reduced to one-fifth its original value. Any other combination between these two limits can be envisioned.

The utility of high permeability fracturing is, thus, compelling—not just for production rate improvement, but also for the remedy of undesirable drawdown-dependent phenomena.

High-Rate Water Packs

Empirical data reported by Tiner et al. (1996), as distilled and presented in Table 5-2, support the frequent notion that high-rate water packs have an advantage over gravel packs, but do not afford the productivity improvement of HPF. This improvement over gravel packs is reasonable by virtue of the additional proppant placed in the perforation tunnels.

While not shown in the table, the performance of these completions over time is also of interest. It is commonly reported that production from high-rate water packs (as in the case of gravel packs) deteriorates with time. By contrast, Stewart et al. (1995), Mathur et al. (1995), and Ning et al. (1995) all report that production may progressively improve (skin values decrease) during the first several months following a HPF treatment.

PERFORMANCE OF FRACTURED HORIZONTAL WELLS IN HIGH PERMEABILITY FORMATIONS

Two of the most important developments in petroleum production in the last 15 years are horizontal wells and high permeability fracturing. Considerable potential is possible by combining the two.

TABLE 5-2.   Skin Values Reported by Tiner et al. (1996)

Gravel Pack	High-Rate Water Pack	HPF
+5 to +10 excellent	+2 to +5 reported	0 to +2 normally
+40 and higher are reported		0 to −3 in some reports

Horizontal wells can be drilled either transverse or longitudinal to the fracture azimuth. The transverse configuration is appropriate for low permeability formations and has been widely used and documented in the literature. The longitudinally fractured horizontal well warrants further attention, specifically in the case of high permeability formations. HPF often results in hydraulic fractures with low dimensionless conductivities. Yet, such fractures installed longitudinally in horizontal wells in high permeability formations can have the net effect of installing a (relative) high conductivity streak in an otherwise limited conductivity flow conduit. Using a generic set of input data, Valkó and Economides (1996) showed discounted revenues for 15 cases that demonstrate this point.

Table 5-3 shows that for a given permeability, the potential for the longitudinally fractured horizontal well is always higher than that of a fractured vertical well and, with realistic fracture widths, may approach the theoretical potential of an infinite conductivity fracture.

Furthermore, the horizontal well fractured with 10-fold less proppant ($C_{fD} = 0.12$) still outperforms the fractured vertical well for $k = 1$ and 10 md, and is competitive at 100 md. The longitudinal configuration may provide the additional benefit of avoiding excess breakdown pressures and tortuosity problems during execution.

DISTINGUISHING FEATURES OF HPF

The Tip Screenout Concept

The critical elements of HPF treatment design, execution, and interpretation are substantially different than for conventional fracture

TABLE 5-3.   Discounted Revenue in US$ (1996) Millions

Configuration	$k = 1$ md	$k = 10$ md	$k = 100$ md
Vertical well	0.73	6.4	57.7
Horizontal well	3.48	14.2	78.8
Fractured vertical well, $C_{fD} = 1.2$	2.59	13.4	89.6
Fractured horizontal well, $C_{fD} = 1.2$	3.88	16.3	95.8
Infinite-conductivity fracture (upper bound for both horizontal and vertical well cases)	3.91	16.3	103.3

treatments. In particular, HPF relies on a carefully timed tip screenout to limit fracture growth and to allow for fracture inflation and packing. This process is illustrated in Figure 5-2.

The TSO occurs when sufficient proppant has concentrated at the leading edge of the fracture to prevent further fracture extension. Once fracture growth has been arrested (and assuming the pump rate is larger than the rate of leakoff to the formation), continued pumping will inflate the fracture (increase fracture width). This TSO and fracture inflation is generally accompanied by an increase in net fracture pressure. Thus, the treatment can be conceptualized in two distinct stages: fracture creation (equivalent to conventional designs) and fracture inflation/packing (after tip screenout).

Figure 5-3 after Roodhart et al. (1994) compares the two-stage HPF process with the conventional single-stage fracturing process. Creation of the fracture and arrest of its growth (tip screenout) is accomplished by injecting a relatively small pad and a 1–4 lbm/gal sand slurry. Once fracture growth has been arrested, further injection builds fracture width and allows injection of higher concentration (e.g., 10–16 lbm/gal) slurry. Final areal proppant concentrations of 20 lbm/ft^2 are possible. The figure also illustrates the common practice of retarding injection rate near the end of the treatment (coincidental with opening the annulus to flow) to dehydrate/pack the well annulus and near-well fracture. Rate reductions may also be used to force tip screenout in cases where no TSO event is observed on the downhole pressure record.

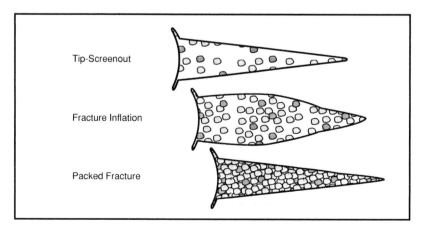

FIGURE 5-2. Width inflation with the tip screenout technique.

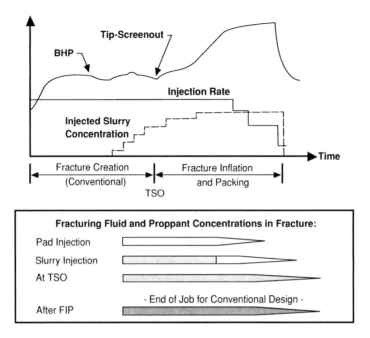

FIGURE 5-3. Comparison of conventional and HPF design concepts.

The tip screenout can be difficult to model, affect, or even detect. There are many reasons for this, including a tendency toward overly conservative design models (resulting in no TSO), partial or multiple tip screenout events, and inadequate pressure monitoring practices.

It is well accepted that accurate bottomhole measurements are imperative for meaningful treatment evaluation. Calculated bottomhole pressures are unreliable because of the dramatic friction pressure effects associated with pumping high sand concentrations through small diameter tubulars and service tool crossovers. Surface data may indicate that a TSO event has occurred when the bottomhole data shows no evidence, and vice versa. Even in the case of downhole pressure data, there has been some discussion of where measurements should be taken. Friction and turbulence concerns have caused at least one operator to conclude that bottomhole pressure data should be collected from below the crossover tool (washpipe gauges) in addition to data collected from the service tool bundle (Mullen et al., 1994).

The detection of tip screenout is discussed further in Chapter 10 along with the introduction of a simple screening tool to evaluate bottomhole data.

Net Pressure and Fluid Leakoff

The entire HPF process is dominated by net pressure and fluid leakoff considerations, first because high permeability formations are typically soft and exhibit low elastic modulus values, and second, because the fluid volumes are relatively small and leakoff rates high (high permeability, compressible reservoir fluids, and non-wall-building fracturing fluids). Also, as described previously, the tip screenout design itself affects the net pressure. While traditional practices applicable to design, execution, and evaluation in MHF continue to be used in HPF, these are frequently not sufficient.

Net Pressure, Closure Pressure, and Width in Soft Formations

Net pressure is the difference between the pressure at any point in the fracture and that of the fracture closure pressure. This definition involves the existence of a unique closure pressure. Whether the closure pressure is a constant property of the formation or depends heavily on the pore pressure (or rather on the disturbance of the pore pressure relative to the long term steady value) is an open question.

In high permeability, soft formations it is difficult (if not impossible) to suggest a simple recipe to determine the closure pressure as classically derived from shut-in pressure decline curves (see Chapter 10). Furthermore, because of the low elastic modulus values, even small, induced uncertainties in the net pressure are amplified into large uncertainties in the calculated fracture width.

Fracture Propagation

Fracture propagation, the availability of sophisticated 3D models notwithstanding, presents complications in high permeability formations, which are generally soft and have low elastic modulus values. For example, Chudnovsky (1996) emphasized the stochastic character of this propagation. Also, because of the low modulus values, an inability to predict net pressure behavior may lead to a significant departure between predicted and actual treatment performance.

It is now a common practice to "predict" fracture propagation and net pressure features using a computer fracture simulator. This trend of substituting clear models and physical assumptions with "knobs"— such as arbitrary stress barriers, friction changes (attributed to erosion, if decreasing, and sand resistance, if increasing) and less-than-well understood properties of the formation expressed as dimensionless "factors"—does not help to clarify the issue.

LEAKOFF MODELS FOR HPF

Considerable effort has been expended on laboratory investigation of the fluid leakoff process for high permeability cores. A comprehensive report can be found in Vitthal and McGowen (1996) and McGowen and Vitthal (1996). The results raise some questions about how effectively fluid leakoff can be limited by filter cake formation.

In all cases, but especially in high permeability formations, the quality of the fracturing fluid is only one of the factors that influence leakoff, and often not the determining one. Transient fluid flow in the formation might have an equal or even larger impact. Transient flow cannot be understood by simply fitting an empirical equation to laboratory data. The use of models based on solutions to the fluid flow equation in porous media is an unavoidable step.

In the following, three models are considered that describe fluid leakoff in the high permeability environment. The traditional Carter leakoff model requires some modification for use in HPF as shown. (Note: While this model continues to be used across the industry, it is not entirely sufficient for the HPF application.) An alternate, filter cake leakoff model has been developed based on the work by Mayerhofer, et al. (1993). The most appropriate leakoff model for high permeability formations may be that of Fan and Economides (1995), which considers the series resistance caused by the filter cake, the polymer-invaded zone, and the reservoir. While the Carter model is in common use, the models of Mayerhofer, et al. and Fan and Economides represent important building blocks and provide a conceptual framework for understanding the key issue of leakoff in high permeability fracturing.

Fluid Leakoff and Spurt Loss as Material Properties: The Carter Leakoff Model with Nolte's Power Law Assumption

There are two main schools of thought concerning leakoff. The first considers the phenomenon as a *material property* of the fluid/rock system. The basic relation (called the integrated Carter equation, given also in Chapter 4) is given in consistent units as

$$\frac{V_L}{A_L} = 2C_L\sqrt{t} + S_p \qquad (5\text{-}2)$$

where A_L is the area and V_L is the total volume lost during the time period from time zero to time t. To make use of material balance, the term V_L must be described. For rigorous theoretical development, V_L is the volume of liquid entering the formation through the two created fracture surfaces of one wing. The integration constant, S_p, is called the spurt loss coefficient and is measured in units of length. It can be considered as the *width* of the fluid body passing through the surface instantaneously at the very beginning of the leakoff process, while $2C_L\sqrt{t}$ is the width of the fluid body following the first slug. The two coefficients, C_L and S_p, can be determined from laboratory or field tests.

As discussed in more detail in Chapter 4, Equation 5-2 can be visualized assuming that the given surface element "remembers" when it has been opened to fluid loss and has its own "zero" time that is likely different from that of other elements along the fracture surface. Points on the fracture face near the well are opened at the beginning of pumping while the points at the fracture tip are younger. Application of Equation 5-2 or its differential form necessitates tracking the opening time for different fracture-face elements, as discussed in Chapter 4.

The second school of thought considers leakoff as a consequence of flow mechanisms in the porous medium, and employs a corresponding mathematical description.

Filter Cake Leakoff Model According to Mayerhofer, et al.

The method of Mayerhofer, et al. (1993) describes the leakoff rate using two parameters that are physically more realistic than the leakoff coefficient: (1) filter cake resistance at a reference time and (2) reservoir permeability. It is assumed that these parameters (R_0, the reference resistance at a reference time t_0, and k_r, the reservoir permeability) have been identified from a minifrac diagnostic test. In addition, reservoir pressure, reservoir fluid viscosity, porosity, and total compressibility are assumed to be known.

Total pressure gradient from inside a created fracture out into the reservoir, Δp, at any time during the injection, can be written as

$$\Delta p(t) = \Delta p_{face}(t) + \Delta p_{piz}(t) + \Delta p_{res}(t) \qquad (5\text{-}3)$$

where Δp_{face} is the pressure drop across the fracture face dominated by the filtercake, Δp_{piz} is the pressure drop across a polymer invaded zone, and Δp_{res} is the pressure drop in the reservoir. This concept is shown in Figure 5-4.

In a series of experimental works using typical hydraulic fracturing fluids (e.g., borate and zirconate crosslinked fluids) and cores with less than 5 md of permeability, no appreciable polymer invaded zone was detected. This simplifying assumption is not valid for linear gels such as HEC (which do not form a filter cake) and may break down for crosslinked fluids at higher permeabilities (e.g., 200 md). Yet, at least for crosslinked fluids in a broad range of applications, the second term in the right-hand side of Equation 4-21 can reasonably be ignored, so

$$\Delta p(t) = \Delta p_{face}(t) + \Delta p_{res}(t) \qquad (5-4)$$

The filter cake pressure term can be expressed as a function of, and is proportional to, R_0, the characteristic resistance of the filter cake. The transient pressure drop in the reservoir can be re-expressed as a series expansion of p_D, a dimensionless pressure function describing the behavior (unit response) of the reservoir. Dimensionless time, t_D, is calculated with the maximum fracture length reached at time

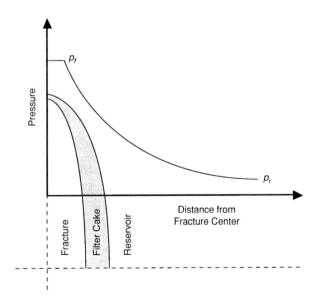

FIGURE 5-4. Filter cake plus reservoir pressure drop in the Mayerhofer et al. (1993) model.

t_n. And r_p is introduced as the ratio of permeable height to the total height (h_p / h_f).

With rigorous introduction of these variables and considerable rearrangement (not shown), an expression for the leakoff rate can be written that is useful for both hydraulic fracture propagation and fracture-closure modeling:

$$q_n = \frac{\Delta p(t_n) - \frac{\mu_r}{\pi k_r r_p h_f} \left[-q_{n-1} p_D(t_{Dn} - t_{Dn-1}) + \sum_{j=1}^{n-1} (q_j - q_{j-1}) p_D(t_{Dn} - t_{Dj-1}) \right]}{\frac{R_0}{2 r_p A_n} \sqrt{\frac{t_n}{t_e}} + \frac{\mu_r p_D(t_{Dn} - t_{Dn-1})}{\pi k_r r_p h_f}} \tag{5-5}$$

This expression allows for the determination of the leakoff rate at any time instant, t_n, if the total pressure difference between the fracture and the reservoir is known, as well as the *history* of the leakoff process. The dimensionless pressure solution, $p_D(t_{Dn} - t_{Dj-1})$, must be determined with respect to a dimensionless time that takes into account the *actual* fracture length at t_n.

The model can be used to analyze the pressure fall-off subsequent to a fracture injection (minifrac) test, as described by Mayerhofer, et al. (1995). The method requires more input data than the similar analysis based on Carter leakoff, but it offers the distinct advantage of differentiating between the two major factors in the leakoff process, filter cake resistance and reservoir permeability.

Polymer-Invaded Zone Leakoff Model of Fan and Economides

The leakoff model of Fan and Economides (1995) concentrates on the additional resistance created by the polymer-invaded zone.

The total driving force behind fluid leakoff is the pressure difference between the fracture face and the reservoir, $p_{frac} - p_i$, which is equivalent to the sum of three separate pressure drops—across the filter cake, the polymer-invaded zone, and in the reservoir:

$$p_{frac} - p_i = \Delta p_{cake} + \Delta p_{inv} + \Delta p_{res} \tag{5-6}$$

The fracture treating pressure is equivalent to the net pressure plus fracture closure pressure (minimum horizontal stress).

When a non-cake building fluid is used, the pressure drop across the filter cake is negligible. This is the case for many HPF treatments.

The physical model of this situation (i.e., fluid leakoff controlled by polymer invasion and transient reservoir flow) is depicted in Figure 5-5. The polymer invasion is labeled in the figure as region 1, while the region of reservoir fluid compression (transient flow) is denoted as 2.

By employing conservation of mass, a fluid flow equation, and an appropriate equation of state, a mathematical description of this fluid leakoff scenario can be written. As a starting point, Equation 5-7 describes the behavior of a Power law fluid in porous media:

$$\frac{\partial^2 p}{\partial x^2} = \frac{n \phi \mu_{eff} c_t}{k} \left(\frac{1}{u}\right)^{1-n} \frac{\partial p}{\partial t} \tag{5-7}$$

where c_t is the system compressibility, k is the formation permeability, u is the superficial flow rate, n is the fluid flow behavior index, ϕ is the formation porosity, and $\mu_{eff} = \frac{K'}{12}\left(9 + \frac{3}{n}\right)^n (150 k \phi)^{\frac{1-n}{2}}$ is the fluid effective viscosity (K' is the power law fluid consistency index).

Combining the description of the polymer-invaded zone and the reservoir, the total pressure drop is given by Fan and Economides (1995) as

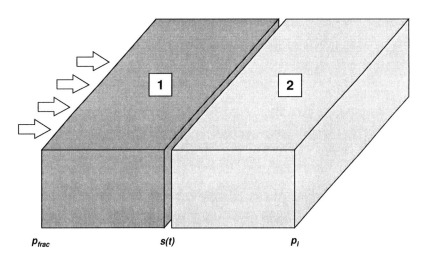

p_{frac} $s(t)$ p_i

FIGURE 5-5. Fluid leakoff model with polymer invasion and transient reservoir flow.

$$p_{frac} - p_r = \frac{\sqrt{\pi}}{2} \frac{\phi\eta}{k} \left\{ \begin{array}{l} \mu_{app} \sqrt{\alpha_1} \; e^{\left(\frac{\eta}{\sqrt{4\alpha_1}}\right)^2} erf\left(\frac{\eta}{\sqrt{4\alpha_1}}\right) \\[2em] + \mu_r \sqrt{\alpha_2} \; e^{\left(\frac{\eta}{\sqrt{4\alpha_2}}\right)^2} erfc\left(\frac{\eta}{\sqrt{4\alpha_2}}\right) \end{array} \right\} \tag{5-8}$$

where $a_1 = \dfrac{k}{n\phi\mu_{eff}\left(\dfrac{1}{u}\right)^{1-n} c_t}$ and $a_2 = \dfrac{k}{\phi\mu c_t}$.

At given conditions, Equation 5-8 can be solved iteratively for the parameter η (not to be confused with fluid efficiency). Once the value of η is found for a specified total pressure drop, the leakoff rate is calculated from

$$q_L = A\left(\frac{\eta}{2\phi}\right)\frac{1}{\sqrt{t}} \tag{5-9}$$

In other words, the factor $\eta/(2\phi)$ can be considered a *pressure-dependent apparent leakoff coefficient.*

FRACTURING HIGH PERMEABILITY GAS CONDENSATE RESERVOIRS

In gas condensate reservoirs, a situation emerges very frequently that is tantamount to fracture face damage. Because of the pressure gradient that is created normal to the fracture, liquid condensate is formed, which has a major impact on the reduction of the relative permeability to gas. Such a reduction depends on the phase behavior of the fluid and the penetration of liquid condensate, which in turn, depends on the pressure drawdown imposed on the well. These phenomena cause an apparent damage that affects the performance of all fractured wells, but especially those with high reservoir permeability.

Wang, et al. (2000) presented a model that predicts the fractured well performance in gas condensate reservoirs, quantifying the effects of gas permeability reduction. Furthermore, they presented fracture treatment design for condensate reservoirs. The distinguishing feature primarily affects the required fracture length to offset the problems associated with the emergence of liquid condensate.

Gas relative permeability curves were derived using a pore-scale network model and are represented by a weighted linear function of immiscible and miscible relative permeability curves:

$$k_{rg} = fk_{rgl} + (1-f)k_{rgM} \qquad (5\text{-}10)$$

where k_{rg} is the gas relative permeability, and f is a weighing factor that is a function of the capillary number,

$$f = \frac{1}{1+\left(\dfrac{N_c}{a}\right)^{1/b}} \qquad (5\text{-}11)$$

The numerical values for a and b are 1.6×10^{-3} and 0.324, respectively, and N_c is the capillary number, defined as

$$N_c = \frac{k\nabla p}{\sigma} \qquad (5\text{-}12)$$

In Equation 5-12, k is the permeability, ∇p is the pressure gradient, and σ is the interfacial tension. The conventional relative permeability for capillary dominated (immiscible) flow in Equation 5-10, k_{rgl}, is defined as

$$k_{rgl} = \left(\frac{S_g}{1-S_{wi}}\right)^{n_g} \qquad (5\text{-}13)$$

where S_g is the gas saturation, S_{wi} is the connate water saturation, and n_g is a constant equal to 5.5. The relative permeability function in the limit of viscous dominated (miscible) flow, k_{rgM}, is defined as

$$k_{rgM} = \frac{S_g}{1-S_{wi}} \qquad (5\text{-}14)$$

Recall that Cinco and Samaniego (1981) provided an expression of the fracture face skin effect that is additive to the dimensionless pressure for the finite conductivity fracture performance:

$$s_{fs} = \frac{\pi b_s}{2x_f}\left(\frac{k}{k_s} - 1\right) \qquad (5\text{-}15)$$

where b_s is the penetration of damage and k_s is the damaged permeability.

An analogy can readily be made for a hydraulically fractured gas condensate reservoir. Liquid condensate that drops out normal to the fracture face can also result in a skin effect, in this case reflecting a reduction in the relative permeability to gas. The penetration of damage would be the zone inside which liquid condensate exists (i.e., the dew point pressure establishes the boundary).

The permeability ratio reduces to the ratio of the relative permeabilities, and because at the boundary k_{rg} is equal to 1, Equation 5-15 becomes simply,

$$s_{fs} = \frac{\pi b_s}{2 x_f} \left(\frac{1}{k_{rg}} - 1 \right)$$

(5-16)

Optimizing Fracture Geometry in Gas Condensate Reservoirs

In gas condensate reservoirs, the fracture performance is likely to be affected greatly by the presence of liquid condensate, tantamount to fracture face damage. An assumption for the evaluation is that the reservoir pressure at the boundary of this "damaged" zone must be exactly equal to the dew point pressure.

For any fracture length and a given flowing bottomhole pressure inside the retrograde condensation zone of a two-phase envelope, the pressure profile normal to the fracture phase and into the reservoir will delineate the points where the pressure is equal to the dew point pressure. From this pressure profile, the distribution of fracture face skin can be determined. The depth of the affected zone is determined from Equation 5-16, the modified Cinco-Ley and Samaniego expression. An additional necessary element is the relative permeability impairment given by the correlation presented in Equations 5-10 to 5-14.

Two example case studies are presented below. The first represents a reservoir with 5 md permeability and a gas condensate with a dew point pressure of 2,545 psi. The flowing bottomhole pressure is 1,800 psi. First, a standard hydraulic fracture optimization—ignoring the effects of the fracture face skin—using a proppant number, N_{prop}, equal to 0.02, results in an expected dimensionless fracture conductivity of 1.6 and a fracture half-length of 220 ft for a 4,000 ft square reservoir. (The value of the proppant number, assuming $k_f = 50,000$ md, $h = 50$ ft, $\rho_p = 165$ lb/ft^3 and $\phi_p = 0.4$, implies a proppant mass approximately equal to 80,000 lb$_m$.) The dimensionless productivity index would be 0.35.

A series of simulations based on the work of Wang, et al. shows the maximum productivity index that can be achieved when the gas condensate skin is introduced, and indicates appropriate changes to the fracture design. The fracture length is progressively increased, while the proppant number (i.e., the mass of proppant injected) is held constant. This, of course, causes an unavoidable reduction in the fracture conductivity, even while maximizing the productivity index.

The results, shown in Figure 5-6, indicate an optimum fracture half-length of 255 ft (16 percent increase from the zero-skin optimum) and an optimum dimensionless conductivity of 1.2 instead of 1.6. Much more significant is the drop in the optimum productivity index to 0.294.

Meeting the expected zero-skin productivity index of 0.35 would necessitate raising the proppant number to approximately 0.045—and more than double the required mass of proppant.

For a much higher permeability reservoir (200 md)—again, ignoring the fracture face skin initially—the same calculation results in an optimum fracture half-length equal to 35 ft (C_{fD} = 1.6). The proppant number for this case is 0.0005 (for the same 80,000 lb_m of proppant). The corresponding dimensionless productivity index is 0.21.

Figure 5-7 is the optimization for the fracture dimensions with gas condensate damage, showing an optimum half-length of 45 ft (a 30

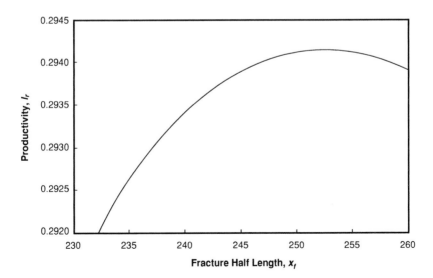

FIGURE 5-6. Optimized fracture geometry in a gas-condensate reservoir (k = 5 md).

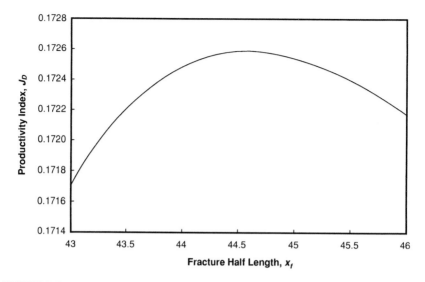

FIGURE 5-7. Optimized fracture geometry in a gas-condensate reservoir (k = 200 md).

percent increase over the zero-skin optimum). The new optimum C_{fD} is 1 and the corresponding productivity index is 0.171.

Here the impact of gas condensate damage on the productivity index expectations and what would be needed to counteract this effect is far more serious. The required proppant number would be 0.003— suggesting 6 times the mass of proppant originally contemplated! In most cases, such a fracture treatment would be highly impracticable, so the expectations for well performance would need to be pared down considerably.

EFFECT OF NON-DARCY FLOW IN THE FRACTURE

Non-Darcy flow is another important issue that deserves specific consideration in the context of HPF. Non-Darcy flow in gas reservoirs causes a reduction of the productivity index by at least two mechanisms. First, the apparent permeability of the formation may be reduced (Wattenbarger and Ramey, 1969) and second, the non-Darcy flow may decrease the conductivity of the fracture (Guppy et al., 1982).

Consider a closed gas reservoir producing under pseudosteady-state conditions, and apply the concept of pseudoskin effect determined by dimensionless fracture conductivity.

Definitions and Assumptions

Gas production is calculated from the pseudosteady-state deliverability equation:

$$q = \frac{\pi k h T_{sc}\left[m(\bar{p}) - m\left(p_{wf}\right)\right]}{p_{sc}T} \times \frac{k_{r,app}}{k_r\left[f_1\left(C_{fD,app}\right) + \ln\left(\frac{0.472r_e}{x_f}\right)\right]} \tag{5-17}$$

where $m(p)$ is the pseudopressure function, $k_{f,app}$ is the apparent permeability of the proppant in the fracture, and $k_{r,app}$ is the apparent permeability of the formation. (All equations in this subsection are given for a consistent system of units, such as SI.) The function f was introduced by Cinco-Ley and Samaniego (1981) and was presented in Chapter 3 as

$$f_1(C_{fD}) = s_f + \ln\frac{x_f}{r_w} = \frac{1.65 - 0.328u + 0.116u^2}{1 + 0.18\ln u + 0.064u^2 + 0.005u^3} \tag{5-18}$$

where $u = \ln C_{fD}$.

The apparent dimensionless fracture conductivity is defined by

$$C_{fD,app} = \frac{k_{f,app}w}{k_{r,app}x_f} \tag{5-19}$$

The apparent permeabilities are flow-rate dependent; therefore, the deliverability equation becomes implicit in the production rate.

Proceeding further requires a model of non-Darcy flow. Almost exclusively, the Forcheimer equation is used:

$$-\frac{dp}{dx} = \frac{\mu}{k}v + \beta\rho|v|v \tag{5-20}$$

where $v = q_a/A$ is the Darcy velocity and β is a property of the porous medium.

A popular correlation was presented by Firoozabadi and Katz (1979) as

$$\beta = \frac{c}{k^{1.2}} \tag{5-21}$$

where $c = 8.4 \times 10^{-8}$ m$^{1.4}$ (= 2.6×10^{10} ft^{-1} md$^{1.2}$).

To apply the Firoozabadi and Katz correlation, we write

$$-\frac{dp}{dx} = \mu v \frac{1}{k}\left(1+\frac{\beta k\rho|v|}{\mu}\right) = \mu v \frac{1}{k}\left(1+\frac{c\rho|v|}{k^{0.2}\mu}\right) \tag{5-22}$$

showing that

$$\frac{k_{app}}{k} = \frac{1}{1+\dfrac{c\rho|v|}{k^{0.2}\mu}} \tag{5-23}$$

The equation above can be used both for the reservoir and for the fracture if correct representative linear velocity is substituted. In the following, it is assumed that $h = h_f$.

A representative linear velocity for the reservoir can be given in terms of the gas production rate as

$$v = \frac{q_a}{4hx_f} \tag{5-24}$$

where q_a is the in-situ (actual) volumetric flow rate; hence, for the reservoir non-Darcy effect,

$$\left(\frac{c\rho v}{k^{0.2}\mu}\right)_r = \left(\frac{c\rho q_a}{2h\mu}\right)\frac{1}{2x_f k_r^{0.2}} \tag{5-25}$$

A representative linear velocity in the fracture can be given in terms of the gas production rate as

$$v = \frac{q_a}{2hw} \tag{5-26}$$

Thus, for the non-Darcy effect in the fracture, one can use

$$\left(\frac{c\rho v}{k^{0.2}\mu}\right)_f = \left(\frac{c\rho q_a}{2h\mu}\right)\frac{1}{wk_f^{0.2}} \tag{5-27}$$

The term ρq_a is the mass flow rate and is the same in the reservoir and in the fracture; $c\rho q_a$ is expressed in terms of the gas production rate as

$$\frac{c\rho q_a}{2h\mu} = \frac{c\rho_a\gamma_g}{2h\mu}q = c_0 q \tag{5-28}$$

where q is the gas production rate in standard volume per time, γ_g is the specific gravity of gas with respect to air, and ρ_a is the density of air at standard conditions. The factor c_0 is constant for a given reservoir-fracture system.

The final form of the apparent permeability dependence on production rate is

$$\left(\frac{k_{app}}{k}\right)_r = \frac{1}{1 + \frac{c_0 q}{2x_f k_r^{0.2}}} \qquad (5\text{-}29)$$

for the reservoir and

$$\left(\frac{k_{app}}{k}\right)_f = \frac{1}{1 + \frac{c_0 q}{w k_f^{0.2}}} \qquad (5\text{-}30)$$

for the fracture. As a consequence, the deliverability equation becomes

$$q = \frac{\pi k h T_{sc}\left[m(\bar{p}) - m(p_{wf})\right]}{p_{sc} T} \times \frac{1}{\left(1 + \frac{c_0 q}{2x_f k_r^{0.2}}\right)\left[f_1\left(C_{fD,app}\right) + \ln\left(\frac{0.472 r_e}{x_f}\right)\right]} \qquad (5\text{-}31)$$

where

$$C_{fD,app} = \frac{k_f w}{k_r x_f} \frac{1 + \frac{c_0}{w k_f^{0.2}} q}{1 + \frac{c_0}{2x_f k_r^{0.2}} q} \qquad (5\text{-}32)$$

The *additional* skin effect, s_{ND}, appearing because of non-Darcy flow, can be expressed as

$$s_{ND} = \left(1 + \frac{c_0 q}{2x_f k_r^{0.2}}\right)\left[f_1\left(C_{fD,app}\right) + \ln\left(\frac{0.472 r_e}{x_f}\right)\right] - \left[f_1\left(C_{fD}\right) + \ln\left(\frac{0.472 r_e}{x_f}\right)\right] \qquad (5\text{-}33)$$

The additional non-Darcy skin effect is always positive and depends on the production rate in a nonlinear manner.

Equations 5-31 and 5-33 are of primary importance to interpret post-fracture well testing data and to forecast production. If the mechanism responsible for the post-treatment skin effect is not understood clearly, the evaluation of the treatment and the production forecast might be severely erroneous.

Case Study for the Effect of Non-Darcy Flow

As previously discussed, non-Darcy flow in a gas reservoir causes a reduction of the productivity index by at least two mechanisms. First, the apparent permeability of the formation may be reduced, and second, the non-Darcy flow may decrease the fracture conductivity. In this case study, the effect of non-Darcy flow on production rates and observed skin effects is investigated.

Reservoir and fracture properties are given in Table 5-4.

A simplified form of Equation 5-30 in field units is

$$q = \frac{\bar{p}^2 - p_{wf}^2}{\dfrac{1424\mu ZT}{k_r h}} \times \frac{1}{\left(1 + \dfrac{c_0 q}{2x_f k_r^{0.2}}\right)\left[f_1\left(C_{fD,app}\right) + \ln\left(\dfrac{0.472 r_e}{x_f}\right)\right]} \tag{5-34}$$

where $c_0 = \dfrac{c\rho_a \gamma_g}{2h\mu}$ must be expressed in ft-md$^{0.2}$/MMSCF/day. In the given example, $c_0 = 73$ ft-md$^{0.2}$/MMSCF/day and

$$C_{fD,app} = \left(\frac{k_f w}{k_r x_f}\right) \frac{1 + \dfrac{c_0}{2x_f k_r^{0.2}} q}{1 + \dfrac{c_0}{w k_f^{0.2}} q} = \left(\frac{k_f w}{k_r x_f}\right) \frac{1 + c_{0r} q}{1 + c_{0f} q} \tag{5-35}$$

where $c_{0r} = 2.34 \times 10^{-3}$ m^3/s $= 7.67 \times 10^{-2}$ (MSCF/day)$^{-1}$
$c_{0f} = 6.14 \times 10^{-1}$ m^3/s $= 2.78 \times 10^{2}$ (MSCF/day)$^{-1}$

Therefore, in field units

$$C_{fD,app} = 1.39\frac{1 + 0.76q}{1 + 280q} \quad \text{and} \tag{5-36}$$

$$q = \frac{4000^2 - p_{wf}^2}{21.645} \times \frac{1}{(1 + 0.76q)\left[f_1\left(C_{fD,app}\right) + 3.16\right]} \tag{5-37}$$

The non-Darcy component of the skin effect can be calculated as

$$s_{ND} = (1 + 0.00076q)\left[f_1\left(C_{fD,app}\right) + 3.16\right] - 4.619 \tag{5-38}$$

The results are shown graphically in Figures 5-8 to 5-10.

TABLE 5-4.   Data for Fractured Well in Gas Reservoir

r_e	ft	1,500
μ	cp	0.02
Z	N/A	0.95
T	°R	640
k_r	md	10
h	ft	80
h_f	ft	80
k_f	md	10,000
x_f	ft	30
w	inch	0.5
γ_g	N/A	0.65
$\bar{p}$	psi	4,000
r_w	ft	0.328

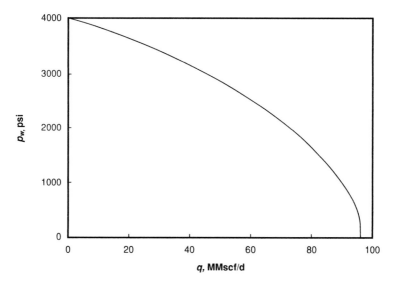

FIGURE 5-8. Inflow performance of fractured gas reservoir, non-Darcy effect from Firoozabadi-Katz correlation.

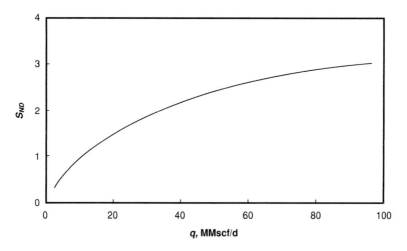

FIGURE 5-9. Additional skin effect from non-Darcy flow in the fracture.

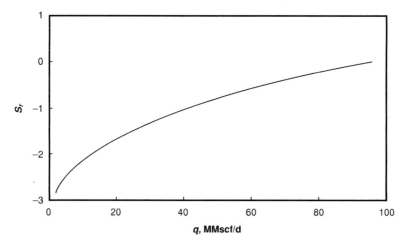

FIGURE 5-10. Observable pseudoskin, the resulting effect of fracture with non-Darcy flow effects.

It is apparent that the effect of the fracture (negative skin on the order of –3) is hidden by the positive skin effect induced by non-Darcy flow. The zero or positive observable skin effect, while directly attributable to the (inevitable) effect of non-Darcy flow, might be interpreted as an unsuccessful HPF job.

Fracturing Materials

Materials used in the fracturing process include fracturing fluids, fluid additives, and proppants. The fluid and additives act together, first to create the hydraulic fracture, and second, to transport the proppant into the fracture. Once the proppant is in place and trapped by the earth stresses ("fracture closure"), the carrier fluid and additives are degraded in-situ and/or flowed back out of the fracture ("fracture cleanup"), establishing the desired highly-productive flow path.

Proppants and chemicals constitute a large share of the total cost to fracture treat a well. The relative value of fracturing materials and pumping costs for treatments performed in the United States are estimated as follows: 45 percent for pumping (pump rental and horsepower charges), 25 percent for proppants, 20 percent for fracturing chemicals, and 10 percent for acid.

Materials and proppants used in hydraulic fracturing have undergone tremendous changes since the first commercial fracturing treatment was performed in 1949 with a few sacks of coarse sand and gelled gasoline as the carrier fluid.

FRACTURING FLUIDS

The fracturing fluid transmits hydraulic pressure from the pumps to the formation, which creates a fracture, and then transports proppant (hence the name *carrier fluid*) into the created fracture. The invasive fluids are then removed (or cleaned up) from the formation, allowing the production of hydrocarbons. Factors to consider when selecting the fluid include availability, safety, ease of mixing and use, viscosity characteristics, compatibility with the formation, ability to be cleaned up from the fracture, and cost.

Fracturing fluids can be categorized as (1) oil- or water-base, usually "crosslinked" to provide the necessary viscosity, (2) mixtures of oil and water, called emulsions, and (3) foamed oil- and water-base systems that contain nitrogen or carbon dioxide gas. Oil-based fluids were used almost exclusively in the 1950s. By the 1990s, more than 90 percent of fracturing fluids were crosslinked water-based systems. Today, nitrogen and carbon dioxide systems in water-based fluids are used in about 25 percent of fracture stimulation jobs.

Table 6-1 lists the most common fracturing fluids in order of current usage. The choice of which crosslinking method to use is based

TABLE 6-1. Crosslinked Fluid Types

Crosslinker	Gelling Agent	pH Range	Application Temperature
B, non-delayed	Guar, HPG	8–12	70–300 °F
B, delayed	Guar, HPG	8–12	70–300 °F
Zr, delayed	Guar	7–10	150–300 °F
Zr, delayed	Guar	5–8	70–250 °F
Zr, delayed	CMHPG, HPG	9–11	200–400 °F
Zr-a, delayed	CMHPG	3–6	70–275 °F
Ti, non-delayed	Guar, HPG, CMHPG	7–9	100–325 °F
Ti, delayed	Guar, HPG, CMHPG	7–9	100–325 °F
Al, delayed	CMHPG	4–6	70–175 °F
Sb, non-delayed	Guar, HPG	3–6	60–120 °F

a—compatible with carbon dioxide

on the capability of a fluid to yield high viscosity while meeting cost and other performance requirements.

Viscosity is perhaps the most important property of a fracturing fluid. Guar gum, produced from the guar plant, is the most common gelling agent used to create this viscosity. Guar derivatives called hydroxypropyl guar (HPG) and carboxymethyl-hydroxypropyl guar (CMHPG) are also used because they provide lower residue, faster hydration, and certain rheological advantages. For example, less gelling agent is required if the guar is crosslinked.

The base guar or guar derivative is reacted with a metal that couples multiple strands of gelling polymer. Crosslinking effectively increases the size of the base guar polymer, increasing the viscosity in the range of shear rates important for fracturing from 5- to 100-fold. Boron (B) is often used as the crosslinking element, followed by organometallic crosslinkers such as zirconium (Zr) and titanium (Ti), and to a lesser extent antimony (Sb) and aluminum (Al).

Foams are especially useful in water-sensitive or depleted (low pressure) reservoirs (Chambers, 1994). Their application minimizes fracture face damage and eases the clean-up of the wellbore after the treatment.

FLUID ADDITIVES

Gelling agent, crosslinker, and pH control (buffer) materials define the specific fluid type and are not considered to be additives. Fluid additives are materials used to produce a specific effect independent of the fluid type. Table 6-2 lists commonly used additives.

Biocides control bacterial contamination. Most waters used to prepare fracturing gels contain bacteria that originate either from contaminated source water or the storage tanks on location. The bacteria produce enzymes that can destroy viscosity very rapidly. Bacteria can be effectively controlled by raising the pH to greater than 12, adding bleach, or employing a broad-spectrum biocide.

Fluid loss control materials provide spurt loss control. The material consists of finely ground particles ranging from 0.1 to 50 microns. The most effective low-cost material is ground silica sand. Starches, gums, resins, and soaps can also be used, with the advantage that they allow some degree of post-treatment cleanup by virtue of their solubility in water. Note that the guar polymer itself eventually controls leakoff, once a filter cake is established.

TABLE 6-2. Fracturing Fluid Additives

Additive	Concentration, gal or lb$_m$ added per 1,000 gallons of clean fluid	Purpose
Biocide	0.1–1.0 gal	Prevents guar polymer decomposition by bacteria
Fluid loss	10–50 lb$_m$	Decreases leakoff of fluid during fracturing
Breakers	0.1–10 lb$_m$	Provides controlled fluid viscosity reduction
Friction reducers	0.1–1.0 gal	Reduces wellbore frictional pressure loss while pumping
Surfactants	0.05–10 gal	Reduces surface tension, prevents emulsions, and changes wettability
Foaming agents	1–10 gal	Provides stable foam with nitrogen and carbon dioxide
Clay control	1–3% KCl typical	Provides temporary or permanent clay (water compatibility)

Breakers reduce viscosity by reducing the size of the guar polymer, thereby having the potential to dramatically improve post-treatment cleanup and production. Table 6-3 summarizes several breaker types and application temperatures.

Surfactants prevent emulsions, lower surface tension, and change wettablilty (i.e., to water wet). Reduction of surface tension allows improved fluid recovery. Surfactants are available in cationic, nonionic, and anionic forms, and are included in most fracturing treatments. Some specialty surfactants provide improved wetting and fluid recovery.

Foaming agents provide the surface-active stabilization required to maintain finely divided gas dispersion in foam fluids. These ionic materials also act as surfactants and emulsifiers. Stable foam cannot be prepared without a surfactant for stabilization.

Clay control additives produce temporary compatibility in water-swelling clays. Solutions containing 1 to 3 percent KCl or other salts are typically employed. Organic chemical substitutes are now available, which are used at lower concentrations.

The type of additives and concentrations used depend greatly on the reservoir temperature, lithology, and fluids. Tailoring of additives for specific applications and advising clients is a main function of the QA/QC chemist.

TABLE 6-3. Fracturing Fluid Breakers

Breaker	Application Temperature	Comments
Enzyme	60–200 °F	Efficient breaker; limit use to pH less than 10
Encapsulated enzyme	60–200 °F	Allows higher concentrations for faster breaks
Persulfates (sodium, ammonium	120–200 °F	Economical; very fast at higher temperatures
Activated persulfates	70–120 °F	Low temperature and high pH applications
Encapsulated persulfates	120–200 °F	Allows higher concentrations for faster breaks
High temperature oxidizers	200–325 °F	Used where persulfates are too fast-acting

PROPPANTS

Because proppants must oppose earth stresses to hold open the fracture after release of the fracturing fluid hydraulic pressure, material strength is of crucial importance. The propping material must be strong enough to bear the closure stress, otherwise the conductivity of the (crushed) proppant bed will be considerably less than the design value (both the width and permeability of the proppant bed decrease). Other factors considered in proppant selection are size, shape, composition, and, to a lesser extent, density.

The two main categories of proppants are naturally occurring sands and manmade ceramic or bauxite proppants. Sands are used for lower-stress applications, in formations approximately 8,000 ft and (preferably, considerably) less. Manmade proppants are used for high-stress situations in formations generally deeper than 8,000 ft. For high permeability fracturing, where a high conductivity is essential, using high-strength proppants may be justified at practically any depth.

There are three primary ways to increase fracture conductivity: (1) increase the proppant concentration, that is, to produce a wider fracture, (2) use a larger (and hence, higher permeability) proppant, or (3) employ a higher-strength proppant, to reduce crushing and improve conductivity. Figures 6-1, 6-2, and 6-3 illustrate the three methods of increasing conductivity through proppant choice.

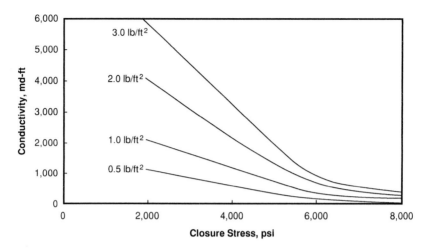

FIGURE 6-1. Fracture conductivity for various areal proppant concentrations (20/40 mesh).

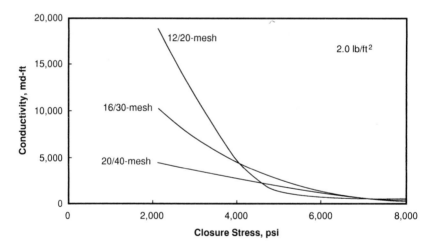

FIGURE 6-2. Fracture conductivity for various mesh sizes.

Figure 6-4 is a selection guide for popular proppant types based on the dominant variable of closure stress.

Calculating Effective Closure Stress

In the course of proppant selection, it is necessary to estimate the magnitude of the closure stress acting on the proppant. The most

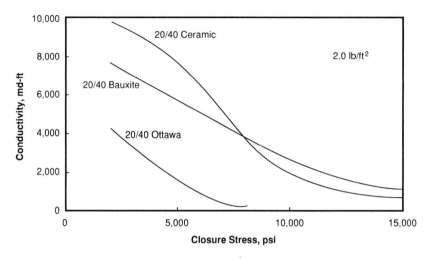

FIGURE 6-3. Fracture conductivity for various proppants.

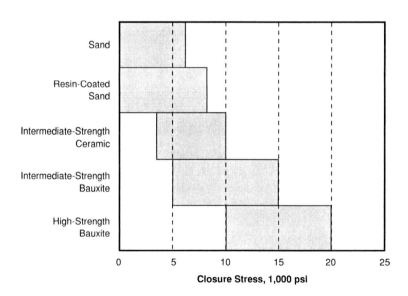

FIGURE 6-4. Proppant selection guide.

common equation used to estimate the closure stress (i.e., the minimum horizontal stress at depth in the reservoir) is known as Eaton's equation. It is commonly given in the form:

$$S_h = \frac{v}{1-v}(S_v - p_p) + p_p \tag{6-1}$$

where v is the Poisson ratio, S_v is the absolute vertical stress, and p_p is the reservoir pore pressure. It is worthwhile to understand the forward development of this relationship.

The absolute vertical stress, S_v, is essentially equal to the force exerted by the weight of the overburden per unit area. Formally, it is the integral of the formation density of the various layers overlying the reservoir. In practice, this value is found to range from 0.95 to 1.1 psi per foot of depth, and in the (typical) absence of specific information, is taken to be equal to 1 psi/ft.

To obtain the effective vertical stress (i.e., the weight of the overburden supported by the rock matrix), the total vertical stress must be reduced by an amount equal to the reservoir pore pressure, giving

$$\sigma_v = S_v - \alpha p_p \tag{6-2}$$

where the coefficient α, called Biot's constant or the *poroelastic* constant, is added to the pore pressure term to account for the fact that reservoir fluids are locally free to move out of the control volume under consideration (not a closed box). This situation is depicted in Figure 6-5.

Biot's constant is typically a value between 0.7 and 1, but most often is taken as unity in order to simplify the already rather approximate calculation.

Now, we know that the *longitudinal strain* that results when a linear elastic solid is placed under a uniaxial load translates to a *lateral strain* according to classic mechanics of materials, that is, the two quantities being related by (in fact defining) the Poisson ratio of the solid, $v = \partial e_x / \partial e_z$. In a similar way, the vertical stress created by the soil layers overlying an oilfield will induce a horizontal stress in the reservoir rock (through the solid matrix). The magnitude of this horizontal stress is calculated by:

$$\sigma_h = \frac{v}{1-v}\sigma_v \tag{6-3}$$

where σ_h is of course the *effective* horizontal stress.

Effective Stress = Total Stress − α (Pore Pressure)

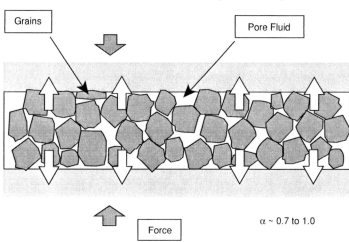

Grains

Pore Fluid

Force

α ~ 0.7 to 1.0

FIGURE 6-5. Poroelasticity.

Combining Equations 6-2 and 6-3 and rearranging slightly,

$$S_h = \frac{\nu}{1-\nu}(S_v - \alpha p_p) + \alpha p_p \qquad (6\text{-}4)$$

which, taking Biot's constant to be equal to 1, yields the common form of Eaton's equation.

Now, it is important to recognize that, unless the producing bottom-hole pressure in the fracture treated well is drawn down to somewhere near zero, the entire burden of this horizontal stress will not be borne by the proppant. Another important observation is that the horizontal stress in the reservoir is itself a function of reservoir pore pressure, so the closure stress on the proppant is nominally reduced with reservoir depletion.

FRACTURE CONDUCTIVITY AND MATERIALS SELECTION IN HPF

Fracture Width as a Design Variable

A great deal has been published concerning optimum fracture dimensions in HPF. While there are debates regarding the optimum, fracture width is largely regarded as more important than fracture length. Of course, this is an intuitive statement and only recognizes the first principle of fracture optimization: higher permeability formations require higher fracture conductivity to maintain an acceptable value of the dimensionless fracture conductivity, C_{fD}.

A "rule of thumb" is that fracture length should be equal to 1/2 of the perforation height (thickness of producing interval). Hunt et al. (1994) showed that cumulative recovery from a well in a 100 md reservoir with a 10 ft damage radius is optimized by extending a fixed 8,000 md-ft conductivity fracture to any appreciable distance beyond the damaged zone. This result implies that there is little benefit to a 50 ft fracture length compared to a 10 ft fracture length. Two observations may be in order. First, the Hunt et al. evaluation is based on cumulative recovery. Second, their assumption of a fixed fracture conductivity implies a decreasing dimensionless fracture conductivity with increasing fracture length (i.e., less than optimal placement of the proppant).

It is generally true that if an acceptable C_{fD} is maintained—this may require an increase in areal proppant concentration from 1.5 lb_m/ft^2, which is common in hard-rock fracturing, to 20 lb_m/ft^2 or more—additional length will provide additional production. As explained in Chapter 3, the optimum fracture conductivity of 1.6 corresponds to the best compromise between the capacity of the fracture to conduct and the capacity of the reservoir to deliver fluids. This applies to high permeability and low permeability formations alike.

The problem, in practice, has been that fracture extent and width are difficult to influence separately. Historically, once a fracturing fluid and injection rate are selected, the fracture width evolves with increasing length according to strict relations (at least in the well-known PKN and KGD design models). Therefore, the key decision variable has been the fracture extent. After the fracture extent is determined, the width is calculated as a consequence of technical limitations (e.g., maximum realizable proppant concentration). Knowledge of the leakoff process helps to determine the necessary pumping time and pad volume.

The tip screenout technique has brought a significant change to this design philosophy. Through TSO, fracture width can be increased without increasing the fracture extent. Now we have a very effective means to design and execute fractures that satisfy the optimum condition.

The ultimate decision requires optimizing the mass of proppant based on economics or, in cases where total fluid and proppant volumes are physically limited (e.g., in offshore environments), optimizing placement of a finite proppant volume

Proppant Selection

The primary and unique issue relating to proppant selection for high permeability fracturing, beyond maintaining a high permeability at any stress, is *proppant sizing*. While specialty proppants such as intermediate strength and resin-coated proppants have certainly been employed in HPF, the majority of treatments are pumped with standard graded-mesh sand.

When selecting a proppant size for HPF, the engineer faces competing priorities: size the proppant to address concerns with sand exclusion, or use maximum proppant size to ensure adequate fracture conductivity.

As with equipment choices and fluids selection, the gravel-packing roots of *frac & pack* are also evident when it comes to proppant selection. Engineers initially focused on sand exclusion and a gravel pack derived sizing criteria such as that proposed by Saucier (1974). Saucier recommends that the mean gravel size (D_{g50}) be five to six times the mean formation grain size (D_{f50}). The so-called "4-by-8 rule" implies minimum and maximum grain-size diameters that are distributed around Saucier's criteria (i.e., $D_{g,min} = 4D_{g50}$ and $D_{g,max} = 8D_{g50}$, respectively). Thus, many early treatments were pumped with standard 40/60 mesh or even 50/70 mesh sand. The somewhat limited conductivity of these gravel pack mesh sizes under in-situ formation stresses is not adequate in many cases. Irrespective of sand mesh size, *frac & packs* tend to reduce concerns with fines migration by virtue of reducing fluid flux at the formation face.

The current trend in proppant selection is to use fracturing-size sand. A typical HPF treatment now employs 20/40 proppant (sand). Maximizing the fracture conductivity can itself help prevent sand production by virtue of reducing drawdown. Results with the larger proppant have been encouraging, both in terms of productivity and limiting or eliminating sand production (Hannah et al., 1993).

It is interesting to note that the topics of formation competence and sanding tendency, major issues in the realm of gravel pack technology, have not been widely studied in the context of HPF. It seems that in many cases HPF is providing a viable solution to completion failures in spite of the industry's primitive understanding of (soft) rock mechanics.

This move away from gravel pack practices toward fracturing-type practices is common to many aspects of HPF with the exception (so far) of downhole tools, and it seems to justify the migration in our terminology from *frac & pack* to high permeability fracturing. The following discussion of fluid selection is consistent with this perspective.

Fluid Selection

Fluid selection for HPF has always been driven by concerns with damaging the high permeability formation, either by filter cake buildup or (especially) polymer invasion. Most early treatments were carried out using HEC, the classic gravel pack fluid, as it was perceived to be less damaging than guar-based fracturing fluids. While the debate has lingered on and while some operators continue to use HEC fluids, the fluid of choice is increasingly borate-crosslinked HPG.

Based on a synthesis of reported findings from several practitioners, Aggour and Economides (1996) provide a rationale to guide fluid selection in HPF. Their findings suggest that if the extent of fracturing fluid invasion is minimized, the degree of damage (i.e., permeability impairment caused by filter cake or polymer invasion) is of secondary importance. They employ the effective skin representation of Mathur et al. (1995) to show that if fluid leakoff penetration is small, even severe permeability impairments can be tolerated without exhibiting positive skin effects. In this case, the obvious recommendation in HPF is to use high polymer concentration, crosslinked fracturing fluids with fluid-loss additives, and an aggressive breaker schedule. The polymer, crosslinker, and fluid-loss additives limit polymer invasion, and the breaker ensures maximum fracture conductivity, a critical factor which cannot be overlooked.

Experimental work corroborates these contentions. Linear gels have been known to penetrate cores of very low permeability (1 md or less) whereas crosslinked polymers are likely to build filter cakes at permeabilities two orders of magnitude higher (Roodhart, 1985; Mayerhofer et al., 1991). Filter cakes, while they may damage the fracture face, clearly reduce the extent of polymer penetration into the reservoir normal to the fracture face. At extremely high permeabilities, even crosslinked polymer solutions may invade the formation.

Cinco-Ley and Samaniego (1981) and Cinco-Ley et al. (1978) described the performance of finite-conductivity fractures and delineated three major types of damage affecting this performance.

■ *Reduction of proppant pack permeability* resulting from either proppant crushing or (especially) unbroken polymer chains leads to fracture conductivity impairment. This can be particularly problematic in moderate to high permeability reservoirs. Extensive progress in breaker technology has dramatically reduced concerns with this type of damage.

■ *Choke damage* refers to the near-well zone of the fracture, which can be accounted for by a skin effect. This damage can result from either over-displacement at the end of a treatment or by fines migration during production. In the latter case, one can envision fines from the formation or proppant accumulating near the well but within the fracture.

■ *Fracture face damage* implies permeability reduction normal to the fracture face, including permeability impairments caused by the filter cake, polymer-invaded, and filter cake-invaded zones.

Composite Skin Effect

Mathur et al. (1995) provide the following representation for effective skin resulting from radial wellbore damage and fracture face damage:

$$s_d = \frac{\pi}{2}\left[\frac{b_2 k_r}{b_1 k_3 + \left(x_f - b_1\right)k_2} + \frac{\left(b_1 - b_2\right)k_r}{b_1 k_1 + \left(x_f - b_1\right)k_r} - \frac{b_1}{x_f} \right] \tag{6-5}$$

Figure 6-6 depicts the two types of damage accounted for in s_d (i.e., fracture-face and radial wellbore damage).

The b- and k- terms are defined graphically in Figure 6-7 and represent the dimensions and permeabilities of various zones included in the finite conductivity fracture model of Mathur et al.

The equivalent damage skin can be added directly to the undamaged Cinco and Samaniego fracture skin effect to obtain the total skin,

$$s_t = s_d + s_f \tag{6-6}$$

Parametric Studies

Aggour and Economides (1996) employed the Mathur et al. model (with no radial wellbore damage) to evaluate total skin and investigate the relative effects of different variables. Their results related the total skin in a number of discrete cases to (1) the depth of fluid invasion normal to the fracture face and (2) the degree of permeability reduction

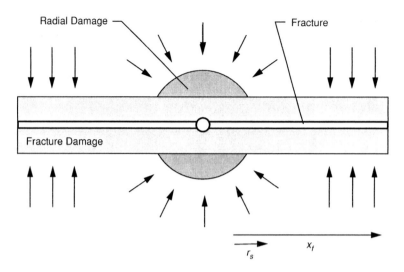

FIGURE 6-6. Fracture face damage.

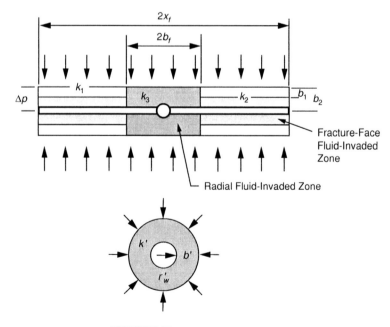

FIGURE 6-7. Fluid invaded zones.

in the polymer-invaded zone. A sample of their results (for $x_f = 25$ ft, $C_{fD} = 0.1$, and $k_f = 10$ md), expressed initially in terms of damage penetration ratios, b_2/x_f, and permeability impairment ratios, k_2/k_r, are re-expressed in real units in Table 6-4. Under each of these conditions, the total skin is equal to zero.

These results suggest that for a (nearly impossible) 2.5 ft penetration of damage, a positive skin is obtained only if the permeability impairment in the invaded zone is more than 90 percent. For a damage penetration of 1.25 ft, the permeability impairment would have to be over 95 percent to achieve positive skins. If the penetration of damage can be limited to 0.25 ft, even a 99 percent permeability reduction in the invaded zone would not result in positive skins. At a higher dimensionless conductivity equal to 1, even higher permeability impairments can be tolerated without suffering positive skins. Thus, if the fracturing fluid leads to a clean and wide proppant pack, penetration and damage to the reservoir can be tolerated.

It is also clear from this work that the extent of damage normal to the fracture face is more important than the degree of damage. If fluid invasion can be minimized, even 99 percent damage can be tolerated. The importance of maximizing C_{fD} is also illustrated; certainly, a good proppant pack should not be sacrificed in an attempt to minimize the fracture face damage.

This points toward the selection of appropriate fracturing fluids:

■ Linear gels by virtue of their considerable leakoff penetration are not recommended.

■ Crosslinked polymer fluids with high gel loadings appear to be much more appropriate.

■ Aggressive breaker schedules are imperative.

■ Filter cake building additives may also be considered to minimize the spurt loss and total leakoff.

TABLE 6-4. Fluid Invasion Damage Tolerated for Zero Skin

Depth of Fluid Invasion Normal to Fracture Face	Permeability Reduction in Invaded Zone
2.5 ft	90%
1.25 ft	95%
0.25 ft	99%

Source: Aggour and Economides (1996).

Work by Mathur et al. (1995) and Ning et al. (1995) further support the conclusion that fracture face damage should not significantly alter long-term HPF performance. The Mathur et al. study of Gulf Coast wells assumed a linear cleanup of the fracture and observed an improvement of the production rate at early time. The Ning et al. study of gas wells in Alberta, Canada, showed that fracture conductivity has the greatest effect on long-term production rates, whereas the effects of polymer invasion were minimal.

Experiments in Fracturing Fluid Penetration

McGowen et al. (1993) presented a series of experiments showing the extent of fracturing fluid penetration in cores of various permeabilities. Fracturing fluids used were 70 lb_m per 1,000-gal HEC and 30 or 40 lb_m per 1,000 gal borate-crosslinked HPG. Filtrate volumes were measured in ml per cm^2 of leakoff area (i.e., cm of penetration) for a 10 md limestone and 200 and 1,000 md sandstones at 120°F and 180°F.

Several conclusions can be drawn from the work:

- Crosslinked fracturing fluids are far superior to linear gels in controlling fluid leakoff. For example, 40 lb_m per 1,000 gal borate-crosslinked HPG greatly outperforms 70 lb_m per 1,000 gal HEC in 200 md core at 180°F.

- Linear gel performs satisfactorily in 10 md rock but fails dramatically at 200 md. Even aggressive use of fluid loss additives (e.g., 40 lb_m per 1,000 gal silica flour) does not appreciably alter the leakoff performance of HEC in 200 md core.

- Increasing crosslinked gel concentrations from 30 to 40 lb_m per 1,000 gal has a major impact on reducing leakoff in 200 md core. Crosslinked borate maintains excellent fluid loss control in 200 md sandstone and performs satisfactorily even at 1,000 md.

This experimental work strongly corroborates the modeling results of Aggour and Economides (1996) and points toward the use of higher-concentration crosslinked polymer fluids with, of course, an appropriately designed breaker system.

Viscoelastic Carrier Fluids

HEC and borate-crosslinked HPG fluids are the dominant fluids currently employed in HPF. However, there is a third class of fluid that deserves to be mentioned, the so-called viscoelastic surfactant, or VES fluids. There is little debate that these fluids exhibit excellent rheological properties and are non-damaging, even in high permeability formations. The elegance of VES fluids is that they do not require the use of chemical breaker additives; the viscosity of this fluid conveniently breaks (leaving considerably less residue than polymer-based fluids) either when it contacts formation oil or condensate or when its salt concentration is reduced. Brown et al. (1996) present typical VES fluid performance data and case histories.

The vulnerability of VES fluids is in their temperature limitation and much higher costs per unit volume. The maximum application temperature for VES fluids has only recently been extended from 130°F up to 240°F.

VES fluids have great potential when considered in a holistic manner: treatments may cost more than polymer fluids, but the resulting appropriately sized fracture could be a far superior producer.

Fracture Treatment Design

Fracture treatment goes well beyond the sizing of a fracture, as important as that is for production enhancement, to include the calculation of a pumping schedule that will realize the goals set for the treatment. This chapter also includes discussion of pre-treatment diagnostics that are often incorporated with fracture treatments to determine or at least place bounds on parameters that are critical to the design procedure and execution.

MICROFRACTURE TESTS

The microfracture stress test ("microfrac") determines the magnitude of the minimum principal in-situ stress of a target formation. The test usually involves the injection of pressurized fluid into a small, isolated zone (4 to 15 ft, 1.2 to 4.6 m) at low injection rates (1 to 25 gal/min, 0.010 to 0.095 m^3/min). The minimum principal in-situ stress can be determined from the pressure decline after shut-in or the pressure increase at the beginning of an injection cycle. The fracture closure pressure and fracture reopening pressure provide good approximations for the minimum principal in-situ stress.

MINIFRACS

The most important test on location before the main treatment is known as a "minifrac," or a fracture calibration test. The minifrac is a pump-in/shut-in test that employs full-scale pump rates and relatively large fluid volumes, on the order of thousands of gallons. Information gathered from a minifrac includes the closure pressure, p_c, net pressure, entry conditions (perforation and near-wellbore friction), and possibly evidence of fracture height containment. The falloff portion of the pressure curve is used to obtain the leakoff coefficient for a given fracture geometry. Figure 7-1 illustrates the strategic locations on a typical pressure response curve registered during the calibration activities.

A minifrac design should be performed along with the initial treatment design. The design goal for the minifrac is to be as representative as possible of the main treatment. To achieve this objective, sufficient geometry should be created to reflect the fracture geometry of the main treatment and to obtain an observable closure pressure from the pressure decline curve. The most representative minifrac would have an injection rate and fluid volume equal to the main treatment, but this is often not practical. In reality, several conflicting design criteria must be balanced, including minifrac volume, created

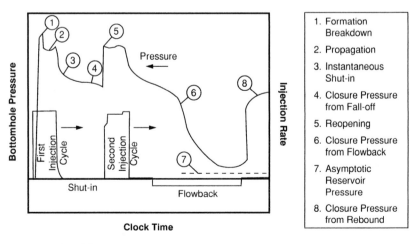

FIGURE 7-1. Key elements on minfrac pressure response curve.

fracture geometry, damage to the formation, a reasonable closure time, and the cost of materials and personnel.

Fracture closure is typically determined from one or more constructions of the pressure decline curve while taking into consideration any available prior knowledge (e.g., that obtained from microfrac tests). Some of the most popular plots used to identify fracture closure pressure are:

- $p_{shut-in}$ vs. t

- $p_{shut-in}$ vs. $\sqrt{t}$

- $p_{shut-in}$ vs. g-function (and variations)

- $\log(p_{ISIP} - p_{shut-in})$

The origin and use of these various plots is sometimes more intuitive than theoretical, which can lead to spurious results. The theoretical basis and limitations of pressure decline analysis must be understood in the context of individual applications. An added complication is that temperature and compressibility effects may cause pressure deviations. In this case, temperature-corrected decline curves can be generated to permit the normal interpretations of the different plot types (Soliman, 1984).

The original concept of pressure decline analysis is based on the observation that the rate of pressure decline during the closure process contains useful information on the intensity of the leakoff process (Nolte, 1979, Soliman and Daneshy, 1991). This stands in contrast to the pumping period, when the pressure is affected by many other factors.

If we assume that the fracture area has evolved with a constant exponent α and remains constant after the pumps are stopped, at time $(t_e + \Delta t)$ the volume of the fracture is given by

$$V_{t_e + \Delta t} = V_i - 2A_e S_p - 2A_e g(\Delta t_D, \alpha) C_L \sqrt{t_e} \qquad (7\text{-}1)$$

where the dimensionless delta time is defined as

$$\Delta t_D = \Delta t / t_e \qquad (7\text{-}2)$$

and the two-variable function $g(\Delta t_D, \alpha)$ can be obtained by integration. Its general form is given by (Valkó and Economides, 1995):

$$g\left(\Delta t_D, \alpha\right) = \frac{4\alpha\sqrt{\Delta t_D} + 2\sqrt{1 + \Delta t_D} \times F\left[\frac{1}{2}, \alpha; 1 + \alpha; \left(1 + \Delta t_D\right)^{-1}\right]}{1 + 2\alpha} \quad (7\text{-}3)$$

The function $F[a, b; c; z]$ is the "Hypergeometric function" available in the form of tables or computing algorithms. For computational purposes (e.g., the included MF Excel spreadsheet for minifrac analysis), the g-function approximations given in Table 7-1 are useful.

Dividing Equation 7-1 by the area, the fracture width at time Δt after the end of pumping is given by

$$\overline{w}_{t_e + \Delta t} = \frac{V_i}{A_e} - 2S_p - 2C_L\sqrt{t_e}\, g\left(\Delta t_D, \alpha\right) \quad (7\text{-}4)$$

Hence, the time variation of the width is determined by the $g(\Delta t_D, \alpha)$ function, the length of the injection period, and the leakoff coefficient, but is not affected by the fracture area.

The decrease of average width cannot be observed directly, but the net pressure during closure is already directly proportional to the average width according to

$$p_{net} = S_f \overline{w} \quad (7\text{-}5)$$

simply because the formation is described by linear elasticity theory (i.e., Equation 4-2). The coefficient S_f is the *fracture stiffness*, expressed in Pa/m (psi/ft). Its inverse, $1/S_f$, is called the *fracture compliance*. For the basic fracture geometries, expressions of the fracture stiffness are given in Table 7-2.

TABLE 7-1.   Approximation of the g-Function for Various Exponents α

$$g\left(d, \frac{4}{5}\right) = \frac{1.41495 + 79.4125\, d + 632.457\, d^2 + 1293.07\, d^3 + 763.19\, d^4 + 94.0367\, d^5}{1. + 54.8534\, d^2 + 383.11\, d^3 + 540.342\, d^4 + 167.741\, d^5 + 6.49129\, d^6}$$

$$g\left(d, \frac{2}{3}\right) = \frac{1.47835 + 81.9445\, d + 635.354\, d^2 + 1251.53\, d^3 + 717.71\, d^4 + 86.843\, d^5}{1. + 54.2865\, d + 372.4\, d^2 + 512.374\, d^3 + 156.031\, d^4 + 5.95955\, d^5 - 0.0696905\, d^6}$$

$$g\left(d, \frac{8}{9}\right) = \frac{1.37689 + 77.8604\, d + 630.24\, d^2 + 1317.36\, d^3 + 790.7\, d^4 + 98.4497\, d^5}{1. + 55.1925\, d + 389.537\, d^2 + 557.22\, d^3 + 174.89\, d^4 + 6.8188\, d^5 - 0.0808317\, d^6}$$

TABLE 7-2. **Proportionality Constant, S_f and Suggested α for Basic Fracture Geometries**

	PKN	KGD	Radial
S_f	$\dfrac{2E'}{\pi h_f}$	$\dfrac{E'}{\pi x_f}$	$\dfrac{3\pi E'}{16 R_f}$
α	4/5	2/3	8/9

The combination of Equations 7-4 and 7-5 yields the following (Nolte, 1979):

$$p = \left(p_C + \frac{S_f V_i}{A_e} - 2S_f S_p \right) - \left(2S_f C_L \sqrt{t_e} \right) \times g(\Delta t_D, \alpha) \tag{7-6}$$

Equation 7-6 shows that the pressure falloff in the shut-in period will follow a straight line trend,

$$p = b_N - m_N \times g(\Delta t_D, \alpha) \tag{7-7}$$

if plotted against the g-function (i.e., transformed time, Castillo, 1987). The g-function values should be generated with the exponent α considered valid for the given model. The slope of the straight line, m_N, is related to the unknown leakoff coefficient by

$$C_L = \frac{-m_N}{2\sqrt{t_e} s_f} \tag{7-8}$$

Substituting the relevant expression for fracture stiffness, the leakoff coefficient can be estimated as given in Table 7-3. This table shows that the estimated leakoff coefficient for the PKN geometry does not depend on unknown quantities because the pumping time, fracture height, and plain strain modulus are assumed to be known. For the other two geometries considered, the procedure results in an estimate of the leakoff coefficient that is strongly dependent on the fracture extent (x_f or R_f).

From Equation 7-6 we see that the effect of the spurt loss is concentrated in the intercept of the straight line with the $g = 0$ axis:

$$S_p = \frac{V_i}{2A_e} - \frac{b_N - p_C}{2S_f} \tag{7-9}$$

TABLE 7-3. **Leakoff Coefficient and No-Spurt Fracture Extent for Various Fracture Geometries**

	PKN	KGD	Radial
Leakoff coefficient, C_L	$\dfrac{\pi h_f}{4\sqrt{t_e}\,E'}(-m_N)$	$\dfrac{\pi x_f}{2\sqrt{t_e}\,E'}(-m_N)$	$\dfrac{8R_f}{3\pi\sqrt{t_e}\,E'}(-m_N)$
Fracture Extent	$x_f = \dfrac{2E'V_i}{\pi h_f^2(b_N - p_C)}$	$x_f = \sqrt{\dfrac{E'V_i}{\pi h_f(b_N - p_C)}}$	$R_f = \sqrt[3]{\dfrac{3E'V_i}{8(b_N - p_C)}}$

As suggested by Shlyapobersky (1987), Equation 7-9 can be used to obtain the unknown fracture extent if we assume there is no spurt loss. The second row of Table 7-3 shows the estimated fracture extent for the three basic models. Note that the no-spurt-loss assumption results in an estimated fracture length for the PKN geometry, but this value is not used to obtain the leakoff coefficient. For the KGD and radial models, fracture extent is calculated first and then used to interpret the slope (i.e., to determine C_L). Once the fracture extent and the leakoff coefficient are known, the lost width at the end of pumping can be easily obtained from

$$w_{Le} = 2g_0(\alpha)C_L\sqrt{t_e} \tag{7-10}$$

The fracture width is

$$\overline{w}_e = \frac{V_i}{x_f h_f} - w_{Le} \tag{7-11}$$

for the two rectangular models and

$$\overline{w}_e = \frac{V_i}{R_f^2 \pi / 2} - w_{Le} \tag{7-12}$$

for the radial model.

Often the fluid efficiency is also determined:

$$\eta_e = \frac{\overline{w}_e}{\overline{w}_e + w_{Le}} \tag{7-13}$$

Note that the fracture extent and the efficiency are *state variables*, which is to say that they will have different values in the minifrac and main treatment. Only the leakoff coefficient is a *model parameter* that can be transferred from the minifrac to main treatment, but even then some caution is needed in its interpretation. The bulk leakoff coefficient determined from the above method is "apparent" with respect to the fracture area. If we have information on the permeable height, h_p, and it indicates that only part of the fracture area falls into the permeable layer, the apparent leakoff coefficient should be converted into a "true" value that corresponds to the permeable area only. This is done by simply dividing the apparent value by r_p (see Equation 7-14).

While adequate for many low permeability treatments, the outlined procedure might be misleading for higher permeability reservoirs. The conventional minifrac interpretation determines a single effective fluid loss coefficient, which usually slightly overestimates the fluid loss when extrapolated to the full job volume (Figure 7-2).

This overestimation typically provides an extra factor of safety in low permeability formations to prevent a screenout. However, this same technique applied in high permeability, or when the differential pressure between the fracture and the formation is high, can significantly overestimate the fluid loss for wall-building fluids (Figure 7-3, Dusterhoft, 1995).

Overestimating fluid leakoff can be highly detrimental when the objective is to achieve a carefully timed tip screenout. In this case,

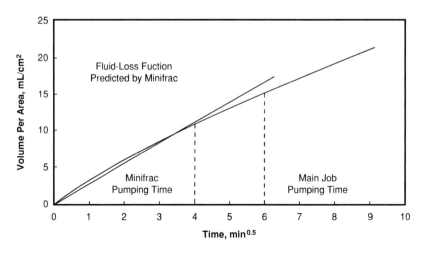

FIGURE 7-2. Fluid leakoff extrapolated to full job volume, low permeability.

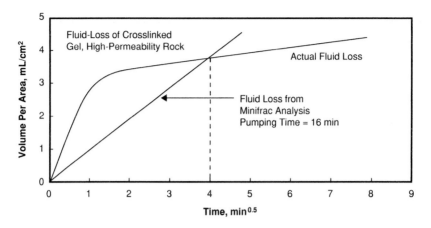

FIGURE 7-3. Overestimation of fluid leakoff extrapolated to full job volume, high permeability.

modeling both the spurt loss and the combined fluid loss coefficient by performing a net pressure match in a 3D simulator is an alternative to classical falloff analysis. This approach is illustrated in Figure 7-4.

Note that the incorporation of more than one leakoff parameter (and other adjustable variables) increases the *degrees of freedom.*

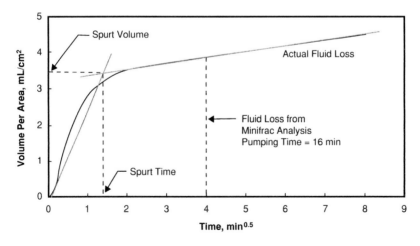

FIGURE 7-4. Leakoff estimate based on a net-pressure match in a 3D simulator (*Source:* Dusterhot et al., 1995).

While a better match of the observed pressure can usually be achieved, the solution often becomes *non-unique* (i.e., other values of the same parameters may provide a similar fit).

TREATMENT DESIGN BASED ON THE UNIFIED APPROACH

We ended Chapter 3 by delineating a certain design logic: for a given amount of proppant reaching the pay layer, we can determine the optimum length (and width). One of the main results was that, for low or moderate proppant numbers (relatively low proppant volumes and/or moderate-to-high formation permeabilities), the optimal compromise occurs at $C_{fD} = 1.6$.

When the formation permeability is above 50 md, it is practically impossible to achieve a proppant number larger than 0.1. Typical proppant numbers for HPF range from 0.0001 to 0.01. Thus, for moderate and high permeability formations, the optimum dimensionless fracture conductivity is always $C_{fDopt} = 1.6$.

In "tight gas" it is possible to achieve large dimensionless proppant numbers, at least in principle. If we assume a limited drainage area and do not question whether the proppant actually reaches the pay layer, a dimensionless proppant number equal to 1 or even 5 can be *calculated*. However, proppant numbers larger than one are not likely in practice.

When the propped volume becomes very large, the optimal compromise happens at larger dimensionless fracture conductivities simply because the fracture penetration ratio cannot exceed unity (i.e., fracture length becomes constrained by the well spacing or limits of the reservoir).

A crucial issue in the design is the assumed fracture height. The relation of fracture height to pay thickness determines the volumetric proppant efficiency. The actual proppant number depends on that part of the proppant that is placed into the pay. It is calculated as the volume of injected proppant multiplied by the volumetric proppant efficiency. Therefore, strictly speaking, an optimum target length can be obtained only if the fracture height is already known. In the following, we assume that the fracture height is known. Later we will return to this issue.

Pump Time

Armed with a target length and assuming that h_f, E', q_i, μ, C_L, and S_p are known, we can design a fracture treatment. The first problem is to determine the pumping time, t_e, using the combination of a width equation and material balance. The first part of a typical design procedure is shown in Table 7-4. Notice that the injection rate, q_i, refers to the slurry (not clean fluid) injected into one wing.

Techniques used to refine K_L are delineated in Tables 7-5 to 7-7.

If the permeable height, h_p, is less than the fracture height, it is convenient to use exactly the same method, but with "apparent" leakoff and spurt loss coefficients. The apparent leakoff coefficient is the "true" leakoff coefficient (the value with respect to the permeable layer) multiplied by the factor r_p, defined as the ratio of permeable to fracture surface (cf. Figures 7-5 and 7-6).

TABLE 7-4. Determination of the Pumping Time

---

1. Calculate the wellbore width at the end of pumping from the PKN (or any

other) width equation: $w_{w,0} = 3.27 \left(\dfrac{\mu q_i x_f}{E'} \right)^{1/4}$ (or rather the non-Newtonian

form shown later)

---

2. Convert wellbore width into average width: $\overline{w}_e = 0.628 w_{w,0}$

---

3. Assume an opening time distribution factor, $K_L = 1.5$ (techniques to refine this value are described below)

---

4. Solve the following equation for t_e:

$\dfrac{q_i t}{h_f x_f} - 2K_L C_L \sqrt{t} - (\overline{w}_e + 2S_p) = 0$ (Quadratic Equation for $x = \sqrt{t}$)

Selecting $\sqrt{t}$ as the new unknown, a simple quadratic equation must be solved:

$at + b\sqrt{t} + c = 0$ where

$a = \dfrac{q_i}{h_f x_f}$; $b = -2K_L C_L$; $c = -(\overline{w}_e + 2S_p)$

---

5. Calculate injected volume: $V_i = q_i t_e$, and fluid efficiency: $\eta_e = \dfrac{h_f x_f \overline{w}_e}{V_i}$

TABLE 7-5.   Refinement of K_L using the Carter II Equation

Calculate an improved estimate of K_L from:

$$K_L = -\frac{S_p}{C_L\sqrt{t_e}} - \frac{\overline{w}_e}{2C_L\sqrt{t_e}} + \frac{\overline{w}_e}{2\eta_e\,C_L\sqrt{t_e}},$$

where $\eta_e = \dfrac{\overline{w}_e(\overline{w}_e + 2S_p)}{4\pi C_L^2 t_e}\left[\exp(\beta^2)\mathrm{erfc}(\beta) + \dfrac{2\beta}{\sqrt{\pi}} - 1\right]$ and $\beta = \dfrac{2C_L\sqrt{\pi t_e}}{\overline{w}_e + 2S_p}$.

If K_L is near enough to the previous guess, stop; otherwise, iterate by repeating the material balance calculation using the new estimate of K_L.

TABLE 7-6.   Refinement of K_L by Linear Interpolation According to Nolte

Estimate the next K_L from

$$K_L = 1.33\eta_e + 1.57(1 - \eta_e),$$

where $\eta_e = \dfrac{\overline{w}_e x_f h_f}{it_e}$.

If K_L is near enough to the previous guess, stop; otherwise, iterate by repeating the material balance calculation using the new estimate of K_L.

TABLE 7-7. K_L from the α Method

Assume a power law exponent α (Table 7-2) and calculate $K_L = g_0(\alpha)$ using equations in Table 7-1. Use the obtained K_L instead of 1.5 in the material balance. (Note that this is not an iterative process.)

For the PKN and KGD geometries, it is the ratio of permeable to the fracture height,

$$r_p = \frac{h_p}{h_f} \tag{7-14}$$

while for the radial model it is given by

$$r_p = \frac{2}{\pi}\left[x(1-x^2)^{0.5} + \arcsin(x)\right] \quad \text{where} \quad x = \frac{h_p}{2R_f} \tag{7-15}$$

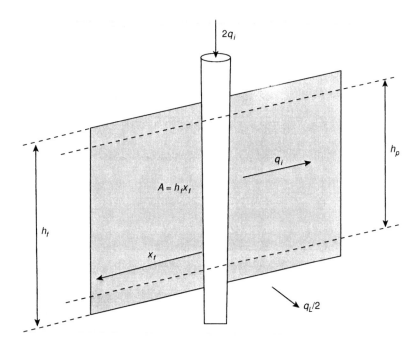

FIGURE 7-5. Ratio of permeable to total surface area, KGD, and PKN geometry.

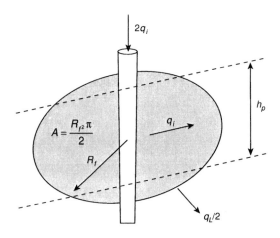

FIGURE 7-6. Ratio of permeable to total surface area, radial geometry.

There are several ways to incorporate non-Newtonian behavior into the width equations. A convenient procedure is to add one additional equation connecting the equivalent Newtonian viscosity with the flow rate. Assuming power law fluid behavior, the equivalent Newtonian viscosity can be calculated for the average cross section using the appropriate entry from Table 4-3. After substituting the equivalent Newtonian viscosity into the PKN width equation, we obtain

$$w_{w,0} = 9.15^{\frac{1}{2n+2}} \times 3.98^{\frac{n}{2n+2}} \left[\frac{1+2.14n}{n} \right]^{\frac{n}{2n+2}} K^{\frac{1}{2n+2}} \left(\frac{i^n h_f^{1-n} x_f}{E'} \right)^{\frac{1}{2n+2}} \quad (7\text{-}16)$$

Proppant Schedule

Given the total pumping time and slurry volume, a stepwise pump schedule (more specifically, a proppant addition schedule, or just *proppant schedule*) is still needed that will yield the designed, propped fracture geometry.

Fluid injected at the beginning of the job without proppant is called the "pad." It initiates and opens up the fracture. Typically, 30 to 60 percent of the fluid pumped during a treatment leaks off into the formation while pumping; the pad provides much of this necessary extra fluid. The pad also generates sufficient fracture length and width to allow proppant placement. Too little pad results in premature bridging of proppant and shorter-that-desired fracture lengths. Too much pad results in excessive fracture height growth and created fracture length. For a fixed slurry volume, excessive pad may result in a final propped length that is considerably shorter than the created (desired) fracture length. Even if the fluid loss were zero, a minimum pad volume would be required to open sufficient fracture width to admit proppant. Generally, a fracture width equal to three times the proppant diameter is felt to be necessary to avoid bridging.

After the specified pad is pumped, the proppant concentration of the injected slurry is ramped up step-by-step until a maximum value is reached at end of the treatment.

Figure 7-7 conceptually illustrates the proppant distribution in the fracture after the first proppant-carrying stage. Most fluid loss occurs in the pad, near the fracture tip. However, some fluid loss occurs along the fracture, and in fact, fluid loss acts to dehydrate the proppant-laden stages. Figure 7-8 shows the concentration of the initial proppant stage climbing from 1 up to 3 lb_m of proppant per gallon of fluid (ppg) as

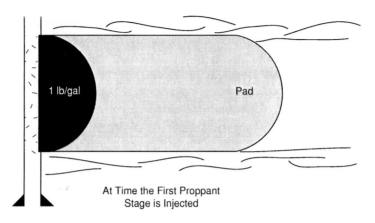

At Time the First Proppant
Stage is Injected

FIGURE 7-7. Beginning of proppant distribution during pumping.

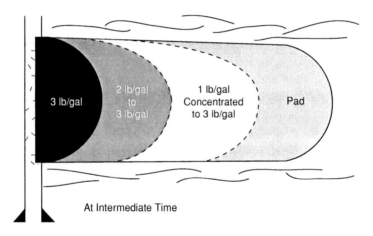

At Intermediate Time

FIGURE 7-8. Evolution of slurry proppant distribution during pumping.

the treatment progresses. Later stages are pumped at higher initial proppant concentrations because they suffer less fluid leakoff (i.e., shorter exposure time and reduced leakoff rates near the well).

Figure 7-9 completes the ideal sequence in which the pad is depleted just as pumping ends and the first proppant stage has concentrated to a final designed value of 5 ppg. The second proppant stage has undergone less dehydration, but also has concentrated to the same

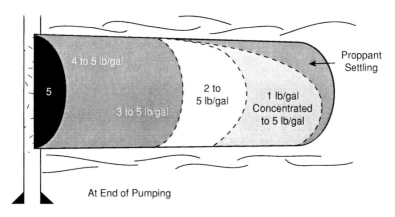

FIGURE 7-9. Proppant concentration in the injected slurry.

final value. If done properly, the entire fracture is filled with a *uniform proppant concentration* at the end of the treatment.

If proppant bridges in the fracture prematurely during pumping, a situation known as a "screen-out," the treating pressure will rise rapidly to the technical limit of the equipment. In this case, pumping must cease immediately (both for the safety of personnel on location and to avoid damaging the equipment), effectively truncating the treatment before the full proppant volume has been placed. Making things worse, the treatment string is often left filled with sand, which then requires incremental rig time and expense to clean out.

TSO designs for highly permeable and soft formations are specifically *intended* to screen out. In this case, it is often possible to continue pumping and inflate the fracture width without exceeding the pressure limits of the equipment because these formations tend to be highly compliant.

While more sophisticated methods are available to calculate the ramped proppant schedule, the simple design technique given in Table 7-8 using material balance and a prescribed functional form (e.g., power law, Nolte 1986) is satisfactory.

One additional parameter must be specified: c_e, the maximum proppant concentration of the injected slurry at the end of pumping. The physical capabilities of the fracturing equipment being used provides one limit to the maximum proppant concentration, but rarely should this be specified as the value for c_e. Ideally, the proppant schedule should be designed to result in a uniform proppant concentration in the fracture at the end of pumping, with the value of the

TABLE 7-8. Proppant Schedule

1. Calculate the exponent of the proppant concentration curve:

$$\varepsilon = \frac{1 - \eta_e}{1 + \eta_e}$$

2. Calculate the pad volume and the time needed to pump it:

$$V_{pad} = \varepsilon V_i \qquad t_{pad} = \varepsilon t_e$$

3. The required proppant concentration (mass per unit of injected slurry volume) curve is given by the following:

$$c = c_e \left(\frac{t - t_{pad}}{t_e - t_{pad}} \right)^{\varepsilon},$$

   where c_e is the maximum end-of-job proppant concentration in the injected slurry.

4. Convert the proppant concentration from *mass per unit of injected slurry volume* into *mass added per unit volume of base fluid* (or "neat" fluid), denoted by c_a, and usually expressed in ppga (pounds added per gallon added of neat fluid).

---

concentration equal to c_e. Therefore, the proppant concentration, c_e, at the end of pumping should be determined from material balance:

$$M = \eta_e c_e V_i \qquad (7\text{-}17)$$

where V_i is the volume of slurry injected in one wing, η_e is the fluid efficiency (or more accurately, slurry efficiency), and M is the mass of injected proppant (one wing).

According to Nolte (1986), the schedule is derived from the requirement that (1) the whole length created should be propped; (2) at the end of pumping, the proppant distribution in the fracture should be uniform; and (3) the proppant schedule should be of the form of a delayed power law with the exponent, ε, and fraction of pad being equal (Table 7-8). More complex proppant scheduling calculations attempt to account for the movement of the proppant both in the lateral and the vertical directions; variations of the viscosity of the slurry with time and location (due to temperature, shear rate and changes in solid content); width requirements for free proppant movement; and other phenomena (Babcock et al. 1967, Daneshy 1974, Shah 1982).

Note that in the above schedule the injection rate q_i refers to the slurry (not clean fluid) injected into one wing. The obtained proppant mass M also refers to one wing.

Continuing our previous example, assume that the target fracture length (152.4 m or 500 ft) was obtained from the requirement to place optimally M = 8,760 kg (19,400 lb$_m$) of proppant into each wing. Using Equation 7-17, we obtain that c_e = 875 kg/m^3 (7.3 lb$_m$/gal). Note that this is still expressed in mass per slurry volume. This means that 12.5 lb$_m$ of proppant must be added to one gallon of neat fracturing fluid (i.e., the added proppant concentration is 12.5 ppga).

The conversion from mass/slurry-volume to mass/neat-fluid-volume is

$$c_a = \frac{c}{1 - \dfrac{c}{\rho_p}} \tag{7-18}$$

where ρ_p is the density of the proppant material.

In our example the fluid efficiency is 19.3 percent, so the proppant exponent and the fraction of pad volume is ε = 0.677. Therefore, the pad injection time is 27.8 min, and after the pad, the proppant concentration of the slurry should be continuously elevated according to the schedule: $c = 875\left(\frac{t-1666}{795}\right)^{0.677}$, where c is in kg/m^3 and t is in seconds, or $c = 7.3\left(\frac{t-27.8}{13.3}\right)^{0.677}$, where c is in lb$_m$/gal of slurry volume and t is in minutes. The obtained proppant curve is shown in Figure 7-10.

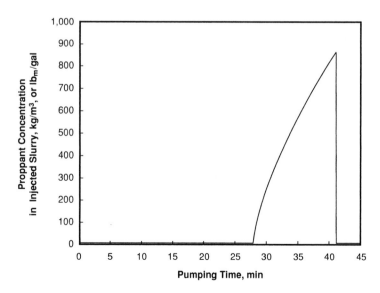

FIGURE 7-10. Evolution of proppant distribution during pumping.

At the end of pumping, the proppant concentration is equal to c_e everywhere in the fracture. Thus, the mass of proppant placed into one wing is $M = V_e \times c_e = \eta_e \times V_i \times c_e$, or in our case, $M = 8{,}760$ kg (19,400 lb$_m$). The average propped width after closure can be determined if the porosity of the proppant bed is known. Assuming $\phi_p = 0.3$, the propped volume is $V_p = M/[(1 - \phi_p)\rho_p]$, or in our case, 6.0 m^3. The average propped width is $w_p = V_p/(x_f \times h_f)$, that is, 2 mm (0.078 in.).

A quick check of the dimensionless fracture conductivity, substituting the propped width, shows that $C_{fD} = (60 \times 10^{-12} \times 0.002)/(5 \times 10^{-16} \times 152) = 1.6$, as it should be for a treatment with a relatively low proppant number.

In the above example, we assumed that the optimum target length and width can be realized without any problem. Of course, it is possible that certain physical or technical constraints (e.g., maximum possible proppant concentration in the slurry) do not allow optimal placement.

Departure from the Theoretical Optimum

In case of conflict, the design engineer has several options. One possibility is to overcome technical limitations by, for example, choosing another type of fluid, proppant and/or equipment.

More often, however, we choose to depart from the theoretical optimum. The art of fracture design is to depart from the theoretical optimum dimensions, but in a reasonable manner and only as much as necessary. In practical terms, this means that the optimum fracture length or pad volume should be reduced or increased by a "factor."

For low permeability formations, the first design attempt often results in very long but narrow fracture. Because there is a certain minimum propped width that is required to maintain continuity of the fracture (e.g., 3 times the proppant diameter), the design engineer should reduce the target length—multiplying it by a factor of 0.5 or sometimes even 0.1. In a careful design procedure, the engineer departs from the theoretical optimum only as much as necessary to satisfy another technical limitation, such as a required minimum width.

In high permeability formations, the first attempt may result in a short fracture with insufficient conductivity (width). This portends a move from the conventional to TSO design, which can produce extremely large fracture widths.

TSO Design

It is the tip screenout or TSO design which clearly differentiates high permeability fracturing from conventional massive hydraulic fracturing. While HPF introduces other identifiable differences (e.g., higher permeability, softer rock, smaller proppant volumes, and so on), it is the tip screenout that makes these fracturing treatments unique. Conventional fracture treatments are designed to propagate laterally and achieve TSO at the end of pumping. In high permeability fracturing, pumping continues beyond the TSO to a second stage of fracture width inflation and packing. It is this two-stage treatment that gives rise to the vernacular of *frac & pack*. The conventional and HPF design concepts were illustrated and compared in Figures 5-3 and 5-4.

Early TSO designs commonly called for 50 percent pad (similar to conventional fracturing) and proppant schedules that ramped-up aggressively; then it became increasingly common to reduce the pad to 10 to 15 percent of the treatment and extend the 0.5 to 2 lb_m/gal stages (which combined may constitute 50 percent of the total slurry volume, for example). Notionally, this was intended to "create width" for the higher concentration proppant addition (e.g., 12 to 14 lb_m/gal).

In our design model (included HF2D Excel spreadsheet), the TSO design procedure differs from the conventional procedure in one basic feature: it uses a "TSO criterion" to separate the lateral fracture propagation period from the width inflation period. This criterion is based on a "dry-to-wet" average width ratio, that is, the ratio of dry width (assuming only the "dry" proppant is left in the fracture) to wet width (dynamically achieved during pumping). According to our assumptions, the screen-out occurs and arrests fracture propagation when the *dry-to-wet width ratio* reaches a critical value.

After the TSO is triggered, injection of additional slurry only serves to inflate the width of the fracture. Thus, it is important to schedule the proppant such that the critical dry-to-wet width ratio is reached at the same time (pumping time) that the *created* fracture length matches the *optimum* fracture length. With the TSO design, practically any width can be achieved—at least in principle. In addition, the first part of any TSO design very much resembles a traditional design, only the target length is reached in a relatively short time, and the dry-to-wet width ratio must reach its critical value during this first part of the treatment.

We suggest a critical dry-to-wet ratio of 0.5 to 0.75 as the TSO criterion (representing quite dehydrated sand in the fracture). Unfortunately, there is no good theoretical or practical method to refine this value. Engineering intuition and previous experience are critical to judging whether a significant arrest of fracture propagation is even possible in a given formation.

There is also no clear procedure to predict if TSO width inflation will be possible in a given formation, though rock mechanics laboratory investigations can suggest the answer. The formation needs to be "soft enough"; in other words, the elasticity modulus cannot be too high. On the other hand, soft formations are often unconsolidated, lacking significant cohesion between the formation grain particles. The main technical limitation to keep in mind is the net pressure, which increases during width inflation. The design engineer should be prepared to depart from the theoretical optimum placement if necessary to keep the fracture treating pressure below critical limits imposed by the equipment.

Another consideration in TSO design is that the created fracture must bypass the assumed damaged region near the wellbore. As such, the design should specify a minimum target length, even if the theoretical optimum calls for a shorter fracture. Often the minimum length is on the order of 50 ft, while the nature of the damage and the length of the perforated interval may dictate other values. Note that this departure from the optimum again can be realized by specifying the "multiply optimum length by a factor" parameter in the provided design software.

PUMPING A TSO TREATMENT

Anecdotal observations related to real-time HPF experiences are abundant in the literature and are not the focus of an engineering-operations text such as this. However, some observations related to treatment execution are in order:

■ Most treatments are pumped using a gravel pack service tool in the "circulate" position with the annulus valve closed at the surface. This allows for live annulus monitoring of bottomhole pressure (annulus pressure + annulus hydrostatic head) and real-time monitoring of the progress of the treatment.

■ When there is no evidence of the planned TSO on the real-time pressure record, the late treatment stages can be pumped at a reduced rate to effect a tip screenout. Obviously, this requires reliable bottomhole pressure data and direct communication by the frac unit operator.

■ Near the end of the treatment, the pump rate is slowed to gravel packing rates and the annulus valve is opened to begin circulating a gravel pack. The reduced pump rate is maintained until tubing pressure reaches an upper limit, signaling that the screen-casing annulus is packed.

■ Because very high proppant concentrations are employed, the sand-laden slurry used to pack the screen-casing annulus must be displaced from surface with clean gel, well before the end of pumping. Thus, proppant addition and slurry volumes must be metered carefully to ensure there is sufficient proppant left in the tubing to place the gravel pack (i.e., to avoid over-displacing proppant into the fracture).

■ Conversely, if an HPF treatment sands out prematurely (i.e., with proppant in the tubing), the service tool can be moved into the "reverse" position and the excess proppant circulated out.

■ Movement of the service tool from the squeeze/circulating position to the reverse position can create a sharp instantaneous drawdown effect and should be done carefully to avoid swabbing unstabilized formation material into the perforation tunnels and annulus.

Swab Effect Example

The following simple equation, given by Mullen et al. (1994) can be used to convert swab volumes into oilfield unit flow rates:

$$q_s = 2,057 \frac{V_s}{t_m} \tag{7-19}$$

where q_s is the instantaneous swab rate in bbl/day, V_s is the swabbed volume of fluid in gal, t_m is the time of tool movement in seconds, and 2,057 is the conversion factor for gal/sec to bbl/day.

The volume of swabbed fluid is calculated from the service tool diameter and the length of stroke during which the sealed service tool does not allow fluid bypass. The average swab volume of a 2.68 in. service tool is 2.8 gal when the service tool is moved from the squeeze position to the reverse-circulation position. Assuming a rather normal movement time of 5 sec, this represents an instantaneous production rate of 1,100 bbl/day.

Perforations for HPF

It is widely agreed that establishing a conductive connection between the fracture and wellbore is critical to the success of HPF, but no consensus or study has emerged that gives definitive direction.

With an eye toward maximizing conductivity and fluid flow rate, many operators shoot the entire target interval with high shot density and large holes (e.g., 12 shots per foot with "big hole" charges). Other operators—more concerned with multiple fracture initiations, near-well tortuosity, and perforations that are not packed with sand—take the extreme opposite approach, perforating just the middle of the target interval with a limited number of 0° or 180° phased perforations. Arguments are made for and against underbalanced versus over-balanced perforating: underbalanced may cause formation failure and "sticking the guns;" overbalanced eliminates a cleanup trip but may negatively impact the completion efficiency.

Solvent or other scouring pills are commonly circulated to the bottom of the workstring and then reversed out to remove scale, pipe dope, or other contaminants prior to pumping into the formation. Several hundred gallons (e.g., 10 to 25 gallons per foot) of 10 to 15 percent HCl acid will then be circulated or bullheaded down to the perforations and be allowed to soak (i.e., to improve communication with the reservoir by cleaning up the perforations and dissolving debris in the perforation tunnel). Some operators are beginning to forego the solvent and acid cleanup (obviously to reduce rig time and associated costs). Their presumption is that the damaging material is pumped deep into the formation and will not seriously impact well performance.

PRE-TREATMENT DIAGNOSTIC TESTS FOR HPF

There are several features unique to high permeability fracturing which make pre-treatment diagnostic tests and well-specific design strategies highly desirable if not essential: fracture design in soft formations is very sensitive to leakoff and net pressure; the controlled nature of the sequential tip screenout/fracture inflation and packing/gravel packing process demands relatively precise execution strategies; and the treatments are very small and typically "one-shot" opportunities. Furthermore, methods used in hard-rock fracturing to determine critical fracture parameters *a priori* (e.g., geologic models, log and core data, or Poisson ratio computational models based on poroelasticity) are of limited value or not yet adapted to the unconsolidated, soft, high permeability formations.

There are three tests (with variations) that form the current basis of pre-treatment testing in high permeability formations: step-rate tests, minifrac tests, and pressure falloff tests.

Step-Rate Tests

The step-rate test (SRT), as implied by the name, involves injecting clean gel at several stabilized rates, beginning at matrix rates and progressing to rates above fracture extension pressure. In a high permeability environment, a test may be conducted at rate steps of 0.5, 1, 2, 4, 8, 10, and 12 barrels per minute, and then at the maximum attainable rate. The injection is held steady at each rate step for a uniform time interval (typically 2 or 3 minutes at each step).

In principle, the test is intended to identify the fracture extension pressure and rate. The stabilized pressure (ideally bottomhole pressure) at each step is classically plotted on a Cartesian graph versus injection rate. Two straight lines are drawn, one through those points that are obviously *below* the fracture extension pressure (dramatic increase in bottomhole pressure with increasing rate), and a second through those points that are clearly *above* the fracture extension pressure (minimal increase in pressure with increasing rate). The point at which the two lines intersect is interpreted as the fracture extension pressure. The dashed lines on Figure 7-11 illustrate this classic approach.

While the conventional SRT is operationally simple and inexpensive, it is not necessarily accurate. A Cartesian plot of bottomhole pressure versus injection rate, in fact, does not generally form a straight line for radial flow in an unfractured well. Simple pressure transient analysis of SRT data using desuperposition techniques shows that with no fracturing the pressure versus rate curve should exhibit upward concavity. Thus, the departure of the real data from ideal behavior may occur at a pressure and rate well below that indicated by the classic intersection of the straight lines (see Figure 7-11).

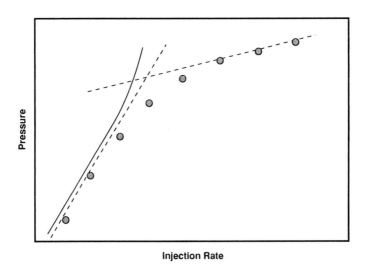

FIGURE 7-11. Ideal SRT—radial flow with no fracturing.

The two-SRT procedure of Singh and Agarwal (1988) is more fundamentally sound. However, given the relatively crude objectives of the SRT in high permeability fracturing, the conventional test procedure and analysis may be sufficient.

The classic test does provide an indication of several things:

- Upper limit for fracture closure pressure (useful in analysis of minifrac pressure falloff data).

- Surface treating pressure that must be sustained during fracturing (or whether sustained fracturing is even possible with a given fluid).

- Reduced rates that will ensure no additional fracture extension and packing of the fracture and near-wellbore with proppant (aided by fluid leakoff).

- Perforation and/or near wellbore friction, which is seldom a problem in soft formations with large perforations and high shot densities.

- Casing pressure that can be expected if the treatment is pumped with the service tool in the circulating position.

A step-down option to the normal SRT is sometimes used specifically to identify near-wellbore restrictions (tortuosity or perforation friction). This test is done immediately following a minifrac or other pump-in stage. By observing bottomhole pressure variations with decreasing rate, near-wellbore restrictions can be immediately detected (i.e., bottomhole pressures that change only gradually as injection rate is reduced sharply in steps is indicative of *no restriction*).

Minifracs

Following the SRT, a minifrac should be performed to tailor the HPF treatment with well-specific information. This is the critical diagnostic test. The minifrac analysis and treatment design modifications can typically be done on-site in less than an hour.

Concurrent with the rise of HPF, minifrac tests, and especially the use of bottomhole pressure information, have become much more common. Otherwise, the classic minifrac procedure and primary outputs as described in the preceding section (i.e., determination of fracture closure pressure and a bulk leakoff coefficient) are widely applied to HPF, this in spite of some rather obvious shortcomings.

The selection of closure pressure, a difficult enough task in hard rock fracturing, can be arbitrary or nearly impossible in high permeability, high-fluid-loss formations. In some cases, the duration of the closure period is so limited (one minute or less) that the pressure signal is masked by transient phenomena. Deviated wellbores and laminated formations (common in offshore U.S. Gulf Coast completions), multiple fracture closures, and other complex features are often evident during the pressure falloff. The softness of these formations (i.e., low elastic modulus) means very subtle fracture closure signatures on the pressure decline curve. Flowbacks are not used to accent closure features because of the high leakoff and concerns with production of unconsolidated formation sand.

New guidelines and diagnostic plots for determining closure pressure in high-permeability formations are being pursued by various practitioners and will eventually emerge to complement or replace the standard analysis and plots.

The shortcomings of classic minifrac analysis are further exposed when used (commonly) to select a single effective fluid loss coefficient for the treatment. As described above, in low permeability formations this approach results in a slight overestimation of fluid loss and actually provides a factor of safety to prevent screenout. In high permeability formations, the classic approach can dramatically underestimate spurt loss (zero spurt loss assumption) and overestimate total fluid loss. This uncertainty in leakoff behavior makes the controlled timing of a tip screenout very difficult. Entirely new procedures based on sound fundamentals of leakoff in HPF (as outlined in Chapter 5) are ultimately needed. The traditional practice of accounting for leakoff with a bulk leakoff coefficient is simply not sufficient for this application.

Pressure Falloff Tests

A third class of pre-treatment diagnostics for HPF has emerged that is *not* common to MHF: pressure falloff tests. Owing to the high formation permeability, common availability of high quality bottomhole pressure data, and multiple pumping and shut-in cycles, matrix formation properties including k_h and skin can be determined from short duration pressure falloff tests using the appropriate transient flow equation. Chapman et al. (1996) and Barree et al. (1996) propose prefrac or matrix injection/falloff tests that involve injecting completion fluid below fracturing rates for a given period of time, and then analyzing the pressure decline using a Horner plot.

The test is performed using standard pumping equipment and poses little interruption to normal operations. A test can normally be completed within one hour or may even make use of data from unplanned injection/shut-in cycles.

The resulting permeability certainly relates to fluid leakoff as described in Chapter 5 and allows the engineer to better anticipate fluid requirements. An initial skin value is useful in "benchmarking" the HPF treatment and for comparison with post-treatment pressure transient analysis.

Bottomhole Pressure Measurements

A discussion of pre-treatment diagnostic tests requires a discussion of the source of pressures used in the analysis. Implicit to the discussion is that the only meaningful pressures are those adjacent to the fracture face, whether measured directly or translated to that point. There are at least four different types of bottomhole pressure data, depending on the location at which the real data are taken:

- Calculated bottomhole pressure—implies bottomhole pressure calculated from surface pumping pressure.

- Deadstring pressure—open annulus, bottomhole pressure deduced knowing density of fluid in annulus; tubing may also be used as dead string when treatment is pumped down the casing.

- Bundle carriers in the workstring—measured downhole, but above the service tool crossover.

- Washpipe data—attached to washpipe below service tool crossover.

Washpipe pressure data is the most desirable for HPF design and analysis based on its location adjacent to the fracture and downstream of all significant flowing pressure drops. Workstring bundle carrier data can introduce serious error in many cases because of fluid friction generated through the crossover tool and in the casing-screen annulus. Without detailed friction pressure corrections that account for specific tool dimensions and annular clearance, there is a possibility for a significant departure between washpipe and workstring bundle carrier pressures. Deadstring pressures are widely used and considered acceptable by most practitioners; some others suggest that redundant washpipe pressure data has shown that the deadstring can mask subtle features of the treatment. The use of bottomhole transducers with real-time surface readouts is suggested in cases where a dead string is not feasible or when well conditions (e.g., transients) may obscure important information.

Reliance on bottomhole pressures calculated from surface pumping pressure is not recommended in HPF. The combination of heavy sand-laden fluids, constantly changing proppant concentrations, very high pump rates, and short pump times makes the estimation of friction pressures nearly impossible.

Fracture Design and Complications

The previous chapter introduced a design procedure that may seem too simplistic for the fracturing engineer. On the surface—considering how many issues are described in the literature or how many phenomena commercial fracture simulators purport to solve—this may seem to be a valid concern.

In this chapter, we show how the concept of *Unified Fracture Design* makes it possible to handle many important topics in a relatively easy manner. Fracture height growth, for example, will ultimately affect the volumetric proppant efficiency (i.e., the proppant number). Proppant embedment into the walls of the created fracture can also be treated as an apparent reduction in the proppant number. Tip effects and non-Darcy flow in the fracture are additional issues that will be addressed.

FRACTURE HEIGHT

Vertical propagation of the fracture is subjected to the same mechanical laws as the lateral propagation. However, if the minimum horizontal stress varies significantly with depth, as it often does, that variation

129

may constrain vertical growth. The equilibrium height concept of Simonson et al. (1978) provides a simple and reasonable method to calculate fracture height when there is a sharp stress contrast between the target layer and the over- and under-burden strata. If the minimum horizontal stress is considerably larger (several hundred psi) in the over- and under-burden layers, we start with a knowledge that the critical stress intensity factor must be exceeded in these adjacent layers before the fracture will grow vertically.

Figure 8-1 illustrates a three layer reservoir system. The middle (pay) layer commonly has the smallest minimum principal stress (σ_1). The two adjacent layers have larger minimum stress. As the pressure at the reference point (center of the perforations) increases, the equilibrium penetrations into the upper (Δh_u) and lower (Δh_d) layers increase. The requirement of equilibrium poses two constraints (one at the top, one at the bottom), resulting in a system of two equations that can be solved simultaneously for the respective dimensionless penetration depths:

$$K_{I,top} = \sqrt{\frac{h_p}{\pi(y_u - y_d)}} \times \int_{-1}^{1} p_n(y)\sqrt{\frac{1+y}{1-y}}\,dy \quad \text{and} \tag{8-1}$$

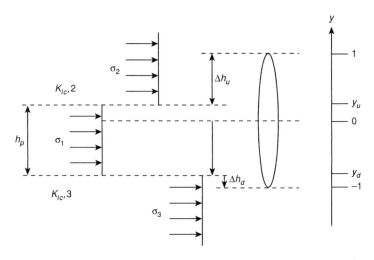

FIGURE 8-1. Notation for fracture height calculations.

$$K_{1,bottom} = \sqrt{\frac{h_p}{\pi(y_u - y_d)}} \times \int_{-1}^{1} p_n(y)\sqrt{\frac{1-y}{1+y}}dy \tag{8-2}$$

where h_p is the pay thickness and the two unknowns are the dimensionless penetration depths y_u and y_d, with values between -1 and $+1$ (see Figure 8-1).

In Equations 8-1 and 8-2, the net pressure is given as a function of the dimensionless vertical location, y, by

$$p_n(y) = k_{00} + k_1 y - \sigma(y) \tag{8-3}$$

$$k_{00} = p_{cp} + \rho g \frac{\Delta h_d - \Delta h_u}{2} \quad \text{and} \tag{8-4}$$

$$k_1 = -\rho g \frac{2h_p}{y_u - y_d} \tag{8-5}$$

where p_{cp} is the pressure at the center of the perforations, ρ is the fluid density, g is the acceleration due to gravity, and $\sigma(y)$ is the minimum stress at the vertical location y.

The solution pair from Equations 8-1 and 8-2 can be used to obtain the upper and lower penetrations in a dimensional length unit (such as ft) according to

$$y_u = 1 - \frac{2\Delta h_u}{h_p + \Delta h_u + \Delta h_d} \quad \text{and} \tag{8-6}$$

$$y_d = -1 + \frac{2\Delta h_d}{h_p + \Delta h_u + \Delta h_d} \tag{8-7}$$

If the hydrostatic pressure component is neglected, the solution is unique up to a certain pressure called the "run-away pressure." Above the run-away pressure, there is no equilibrium state. This does not mean that an unlimited height growth occurs, but rather that there is no reason to believe the vertical growth will be more constrained than the lateral one. As a consequence, we can assume a radially propagating fracture with a circular shape. If the hydrostatic pressure component (due to ρ) is *not* neglected, an interesting phenomenon arises. Once the pressure at the perforations reaches a critical value, another pair of solutions appears. This will be illustrated in the next section.

Fracture Height Map

With the appropriate sample data as provided in Table 8-1 for three adjacent layers, the previous fracture height equations can be used to generate a *fracture height map* as shown in Figure 8-2. The fracturing fluid is water based.

As the bottomhole treating pressure exceeds 20.68 MPa (3000 psi), a fracture is opened. The fracture will penetrate both the overlying and underlying strata, but the penetration into the upper layer is larger because there is less stress contrast to contain the upward growth.

If the hydrostatic pressure component is taken into consideration (theoretical development not shown), there is a certain bottomhole treating pressure at which the original solution set is no longer unique, and a second pair of solutions appears. For our example, this second set of solutions occurs at bottomhole pressures above 25.3 MPa (3,675 psi) as shown using dashed lines in Figure 8-2. The appearance of the second pair of solutions is an indication of *instability*. For safe containment of the created fracture the bottomhole treating pressure should remain below 25.3 MPa (3,675 psi).

As long as the net pressure remains in a safe range, the created fracture height can be read from the fracture height map. Once the anticipated net pressure is large enough to exceed the safe region, or in cases where there is no evidence of a stress contrast, then it is safe to assume a radial extension (i.e., a circular or *penny-shaped* fracture).

TABLE 8-1.   Data for the Fracture Height Map Example

h_p	15.24 m	50 ft
σ_1	20.68 MPa	3,000 psi
σ_2	24.13 MPa	3,500 psi
σ_3	27.58 MPa	4,000 psi
$K_{IC,2}$	1.01 MPa·m$^{1/2}$	1,000 psi·in.$^{1/2}$
$K_{IC,3}$	1.01 MPa·m$^{1/2}$	1,000 psi·in.$^{1/2}$
ρ	1,000 kg/m^3	62.4 lb$_m$/ft^3

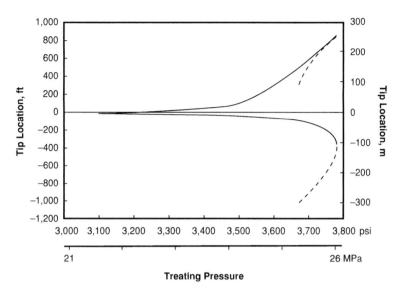

FIGURE 8-2. Fracture height map.

Practical Fracture Height Determination

The equilibrium height concept can be applied in an averaged manner to determine a constant fracture height from the calculated net pressure (Rahim and Holditch, 1993). Because any calculation needs an *a priori* estimate of the fracture height, this can be done in successive iterations. In the so-called pseudo-3D fracture propagation models (Palmer and Caroll, 1983, Settari and Cleary, 1986, Morales and Abou-Sayed, 1989), the equilibrium concept is applied in every time step and at every lateral location.

In conjunction with our unified fracture design procedure, it is often satisfactory to make a preliminary judgment on fracture height growth. If a considerable stress contrast can be anticipated, it is suggested to prepare a height map and use it in an iterative manner to specify the height, using the calculated net pressure as a corrector variable in the iteration process.

However, if there is no evidence of a sharp stress contrast, it is suggested to assume a 1:1 aspect ratio ($h_f = 2x_f$) or a 2:1 aspect ratio ($h_f = x_f$). In practice, this means iteratively modifying the fracture height until the desired ratio is satisfied.

The primary variable responsible for stress contrast is the Poisson ratio. Equation 6-4 (Eaton's equation) can be used to estimate stress contrast from the Poisson ratio. More elaborate correlations are available for certain geographic regions.

TIP EFFECTS

Inherent in the early 2D models was an assumption that there is no net pressure at the tip of the fracture. This *zero net pressure* assumption implies that the energy dissipated during the creation of a new fracture surface can be neglected. Frequent discrepancies between field results and theoretical predictions based on linear elastic fracture mechanics (LEFM) suggest that the zero net pressure assumption cannot be valid in general. Several practitioners cite evidence of "abnormally high" fracturing pressures (Medlin and Fitch, 1988, Palmer and Veatch, 1990), that is, measured field net pressures larger than that predicted by their simulators. Net pressures that are insensitive to rate variations and fluid viscosity are another indication of non-LEFM behavior.

Our present understanding is that in most cases fracture propagation is retarded by "tip effects." This means elevated pressures near the tip of the fracture (indicative of intensive energy dissipation) and a more uniform pressure profile in the main part of the fracture. Several attempts have been made to incorporate this tip phenomenon into fracture propagation models. One reasonable approach is to introduce an *apparent fracture toughness* that increases with the size of the fracture (Shlyapobersky, 1987). Other modelers introduce a controlling relationship for the fracture propagation velocity, u_f, derived from various considerations such as a *fluid lag region* near the fracture tip; non-linear rock deformation and *dilatancy* at the fracture tip; or a complicated rate- and scale-dependent *process zone* (statistical distribution of microcracks).

The continuum damage mechanics (CDM) approach considers fracture propagation as evolution of damage in the material. As the fracture tip approaches a point in the formation, the stress state changes at that point (the fracture acts as a stress concentrator). The increased load causes the evolution of local damage. When the damage reaches a critical value, the location joins the fracture. This reaction of the formation to stress state is described by a combined parameter, Cl^2, which comprises a damage parameter (the Kachanov parameter), C, and a scale parameter, l.

A boundary condition based on the combined CDM parameter is written to replace the zero net pressure boundary condition in a simple 2D differential model (such as PKN), resulting in the following form of the tip propagation velocity:

$$u_f = \frac{C\bar{l}^2}{\pi\sigma_{H,min}} \left(\frac{x_f^{1/2}}{\bar{l} + x_f} \right)^2 w_{x=x_f}^2 \tag{8-8}$$

When the dimensionless version of the CDM parameter, $C_D l_D^2$, is near unity, the fracture evolution is unretarded. When this parameter is on the order of 0.01 or less, the propagation velocity is less than the one calculated from a simple 2D model. The overall effect of a small dimensionless CDM parameter is the increase of net pressure (and corresponding increase of fracture width).

Because the CDM parameter can vary by several orders of magnitude between field observations and laboratory estimates, this additional piece of information is best derived from the fracture propagation pressure observed during a minifrac. The CDM parameter can be adjusted in the appropriate 2D design model until the predicted net pressure matches that observed during the minifrac. This matching process is incorporated automatically in the unified fracture design. (Refer to the PKN-CDM option in the included MF Excel spreadsheet for minifrac evaluation, and the respective design option in the HF2D Excel spreadsheet). The resulting CDM parameter is then used throughout the design procedure.

Additional theoretical background, computational results, and CDM-PKN design examples are provided in *Hydraulic Fracture Mechanics*.

NON-DARCY FLOW IN THE FRACTURE

For high-rate gas wells, where a certain percentage liquid content in the gas is inevitable, the concept of proppant pack permeability deserves special attention. When gas-liquid mixtures flow in a propped fracture with high velocity, the liquid droplets collide with the proppant grains, resulting in a significant dissipation of energy (loss of pressure). Thus, the magnitude of pressure drop in the fracture is greater than would be predicted based on the nominal permeability contrast between the fracture and formation. The fracture behaves with an apparent permeability that is far less than the nominal value measured under single phase flow conditions. There is extensive literature available describing this non-Darcy flow effect in the fracture (Jin and Penny, 2000, M. Cikes, 2000, Milton-Tayler, 1993, Gidley, 1990, Guppy et al., 1982), as well as our treatment of the subject in Chapter 5.

For the current purpose, it is enough to understand that at actual flow conditions the proppant pack can be described by an apparent permeability or, in other terms, the nominal permeability multiplied by a correction factor. Depending on the velocity of the gas, liquid content, droplet size distribution, and proppant quality, the correction factor can be as low as 0.1.

The treatment of non-Darcy flow within unified fracture design is relatively simple. Using an estimated correction factor, the apparent proppant permeability should be reduced, for instance, from 60,000 md to 10,000 md. Of course, this reduces the proppant number and the corresponding maximum productivity index calculated by the fracture design spreadsheet. An anticipated gas velocity can be calculated from the reduced PI (a pressure drawdown must be assumed, and the properties of the gas-liquid mixture must be known). The estimate of gas velocity, in turn, can be used to improve the original estimate of the non-Darcy correction factor. This process can be done iteratively as necessary to arrive at a corrected proppant number for use throughout the design.

Use of a corrected proppant number is illustrated in example MPF03 later in this chapter.

COMPENSATING FOR FRACTURE FACE SKIN

In a certain reservoir, it is suspected that the fracturing fluid filtrate will interact with the formation and create an estimated fracture face skin, $s_{ff} = 1$. What is the effect of this phenomenon on the productivity of the well, and how can we compensate for it? Assume the proppant number for the suggested treatment is $N_{prop} = 0.1$.

Recall that the maximum dimensionless productivity index that can be achieved with $N_{prop} = 0.1$ (see Chapter 3) is

$$J_{Dmax} = \frac{1}{0.99 - 0.5\ln N_{prop}} = 0.47 \qquad (8\text{-}9)$$

If there is a fracture face skin, $s_{ff} = 1$, and we assume the simple case of uniform influx, then the actual productivity will be

$$J_{D\,actual} = \frac{1}{0.99 - 0.5\ln N_{prop} + 1} = 0.32 \tag{8-10}$$

The fracture face skin causes a considerable decrease in productivity. From the equation it is seen that approximately $e^2 = 7.4$ times more proppant would compensate for the loss of productivity caused by a fracture face skin of 1.

EXAMPLES OF PRACTICAL FRACTURE DESIGN

In the remaining part of this chapter, we will illustrate the design logic incorporated in unified fracture design. We intentionally consider cases where only limited data are available.

A Typical Preliminary Design—Medium Permeability Formation: MPF01

Table 8-2 shows available data for a "medium" permeability formation (with a permeability of 1.7 md and net pay of 76 ft). The input data contains the well radius and the drainage radius (calculated from 40 acre spacing). These important reservoir parameters should not be missed.

A preliminary sizing decision is that 90,000 lb_m of proppant should be injected. At the closure stress anticipated (5,000 psi), the selected resin-coated 20/40 mesh sand will have an in-situ permeability of 60,000 md. This number already incorporates the effect of some proppant crushing and the decrease of proppant pack permeability due to imperfect breaking of the gel. Obviously, this is one of the key parameters of the design, and the design engineer must do everything possible to make this estimate as relevant as possible. (It is not enough to purchase an expensive 3D program with vendor-provided proppant data and "click" on the name of the proppant.)

The plane-strain modulus (i.e., basically the Young's modulus) is 2×10^6 psi. Minifrac tests in the same formation with the same fluid usually result in a leakoff coefficient of 0.005 ft/min$^{1/2}$; some spurt loss is also anticipated. (Note that these values are with respect to the pay

TABLE 8-2.   Input Data for MPF01

Proppant mass for (two wings), lb_m	90,000
Sp grav of proppant material (water = 1)	2.65
Porosity of proppant pack	0.38
Proppant pack permeability, md	60,000
Max propp diameter, D_{pmax}, inch	0.031
Formation permeability, md	1.7
Permeable (leakoff) thickness, ft	76
Well radius, ft	0.25
Well drainage radius, ft	745
Pre-treatment skin factor	0.0
Fracture height, ft	
Plane strain modulus, E' (psi)	2.0E + 06
Slurry injection rate (two wings, liq + prop), bpm	20.0
Rheology, K' $(lb_f/ft^2) \times s^{n'}$	0.07
Rheology, n'	0.45
Leakoff coefficient in permeable layer, $ft/min^{1/2}$	0.005
Spurt loss coefficient, S_p, gal/ft^2	0.010

layer. It is assumed that there is no leakoff outside of the pay layer.) The fluid rheology parameters are provided by the service company and (because of pressure limitations in this case) the injection rate is 20 bpm.

Note that the fracture height line is still empty in the summary of input data (Table 8-2). We know that the gross pay (i.e., the distance between the top and bottom of the producing interval) is 100 ft. Within this interval, however, only 76 ft is pay. A preliminary estimate of fracture height should be a minimum of 100 ft, but the actual height will be related to several other factors.

A reasonable assumption, in the absence of any reliable data on stress contrast, is that the aspect ratio of the created fracture is 2:1. In other words, we will find the fracture height, h_f, by adjusting it to the target length, according to $h_f = x_f$.

At this point, we input a starting estimate of $h_f = 100$ ft into our design spreadsheet, and we specify the additional operational constraints as shown in Table 8-3.

TABLE 8-3. Additional Input for MPF01

Max possible added proppant concentration, lb_m/gal neat fluid	12
Multiply opt length by factor	1
Multiply Nolte pad by factor	1

According to the service company, the maximum available proppant concentration is 12 ppga (lb_m proppant added to 1 gallon of neat fracturing fluid). The other two parameters are fixed at their default value.

The output of the first run of our design spreadsheet contains three parts. In the first part, a "wish list" is shown (Table 8-4).

It states that the proppant number is 0.35, and with the proppant placed optimally, we could achieve a dimensionless productivity index of 0.65 and a skin factor as negative as –5.72. The folds of increase in productivity is 4.74 (over the zero skin situation we fixed in the input as the basis of comparison).

However, a warning message (displayed on the screen in red, boldface here) indicates that our wish-list could not be realized:

Suboptimal placement with constraints satisfied

Mass of proppant reduced

The actual placement the design program was able to produce is somewhat disappointing, as shown in Table 8-5.

TABLE 8-4. Theoretical Optimum for MPF01 (h_f = 100 ft)

Output	
Optimum Placement without Constraints	
Proppant number, N_{prop}	0.3552
Dimensionless PI, J_{Dopt}	0.65
Optimal dimensionless fracture cond, C_{fDopt}	1.8
Optimal half length, x_{fopt}, ft	294.2
Optimal propped width, w_{opt}, inch	0.2
Post treatment pseudo-skin factor, s_f	–5.72
Folds of increase of PI	4.74

TABLE 8-5.   Actual Placement for MPF01 (h_f = 100 ft)

Actual placement	
Proppant mass placed (2 wing)	58,501
Proppant number, N_{prop}	0.2309
Dimensionless PI, J_{Dact}	0.57
Dimensionless fracture cond, C_{fD}	1.2
Half length, x_f, ft	294.2
Propped width, w, inch	0.12
Post treatment pseudo-skin factor, s_f	–5.50
Folds of increase of PI	4.15

In other words, the design program can only assure the placement of 58,500 lb$_m$ of proppant. The reason for this will be discussed later. At this point, we should not pay too much attention to it, because our specified fracture height of 100 ft was not realistic.

To approach our desired 2:1 aspect ratio ($h_f = x_f$), we increase the fracture height to 200 ft. The calculated theoretical optimum target length is now $h_f = 216$ ft. A third adjustment to $h_f = 211$ ft will finally establish the desired aspect ratio (Table 8-6).

We see that the proppant number is significantly smaller than it was previously. Why did this happen? Because the increase in fracture

TABLE 8-6.   Theoretical Optimum for MPF01 (h_f = 211 ft)

Output	
Optimum Placement without Constraints	
Proppant number, N_{prop}	0.1684
Dimensionless PI, J_{Dopt}	0.53
Optimal dimensionless fracture cond, C_{fDopt}	1.6
Optimal half length, x_{fopt}, ft	211.1
Optimal propped width, w_{opt}, inch	0.1
Post treatment pseudo-skin factor, s_f	–5.37
Folds of increase of PI	3.85

height decreases the volumetric proppant efficiency (i.e., the part of proppant that is "working for us"). The optimum length corresponding to this proppant number is 211 ft, which means that our fracture, if it can be realized, will have the desired 2:1 aspect ratio. But can it be realized?

According to the red message (shown here in boldface), the optimum placement can be realized:

Constraints allow optimum placement

We found that the 90,000 lb_m proppant can be placed into the well, though not all of the proppant will be placed into the pay layer (Table 8-7).

The part of the proppant reaching the pay will represent a proppant number, N_{prop} = 0.168, and the corresponding optimum half length is 211 ft. The treatment will establish a dimensionless productivity index, J_{Dact} = 0.53. In other words, a negative equivalent skin, s_f = –5.37, will be created.

Note that the entire design logic is based on the proppant number concept. We do not specify an arbitrary length; rather, we obtain the optimum length. The design process makes sure that the desired length is realized and that the desired volume of proppant is placed uniformly.

Some details of the treatment are shown in Table 8-8. More details can be found by running the HF2D Excel spreadsheet.

TABLE 8-7. Actual Placement for MPF01 (h_l = 211 ft)

Actual Placement	
Proppant mass placed (2 wing)	90,000
Proppant number, N_{prop}	0.1684
Dimensionless PI, J_{Dact}	0.53
Dimensionless fracture cond, C_{fD}	1.6
Half length, x_f, ft	211.1
Propped width, w, inch	0.12
Post treatment pseudo-skin factor, s_f	–5.37
Folds of increase of PI	3.85

TABLE 8-8. Details of the Actual Pacement for MPF01 (h_f = 211 ft)

Treatment Details	
Efficiency, e_{ta}, %	34.5
Pumping time, t_e, min	40.4
Pad pumping time, t_e, min	19.7
Exponent of added proppant concentration, e_{ps}	0.4871
Uniform proppant concentration in frac at end, lb_m/ft^3	57.5
Areal proppant concentration after closure, lb_m/ft^2	1.0
Max added proppant concentration, lb per gal clean fluid	11.8
Net pressure at end of pumping, psi	132.5

Pushing the Limit—Medium Permeability Formation: MPF02

For illustrative purposes, we will consider MPF01 as our base case. In this section, we explore whether 150,000 lb_m of proppant can be placed in a similar manner. If so, what good will it do for the well productivity?

The detailed design iteration was illustrated in the previous example, therefore, Table 8-9 shows only the main results. Table 8-10 shows the theoretical optimum for MPF02.

The first thing we should note is that the increase in proppant volume and corresponding larger proppant number will yield only a marginal improvement in productivity, even if everything goes well. This should make us consider whether it is worth "pushing the limit."

TABLE 8-9. Input for MPF02

Proppant mass for (two wings), lb_m	150,000
. . .	
Fracture height, ft	248
. . .	

TABLE 8-10. Theoretical Optimum for MPF02

Output	
Optimum Placement without Constraints	
Proppant number, N_{prop}	0.2387
Dimensionless PI, J_{Dopt}	0.58
Optimal dimensionless fracture cond, C_{fDopt}	1.7
Optimal half length, x_{fopt}, ft	248.0
Optimal propped width, w_{opt}, inch	0.1
Post treatment pseudo-skin factor, s_f	−5.54
Folds of increase of PI	4.23

Even more food for thought is provided by the message:

Suboptimal placement with constraints satisfied

Mass of proppant reduced

and by the next output, given in Table 8-11.

As we see, the design program had to reduce the amount of proppant placed into the formation. With this reduction, the actual folds of increase is hardly more than what we can achieve with 90,000 lb_m of proppant. It is obvious that "pushing the limit" in this case is not worth the effort and money.

But is it really obvious? Several service companies would be happy to supply (at an appropriate premium) better equipment with the capacity to pump proppant concentrations as high as 16 ppga.

So, let's change the operational constraint and run the design again!

Max possible added proppant concentration, lb_m/gal neat fluid	16

The message is now encouraging (Table 8-12):

Constraints allow optimum placement

TABLE 8-11.   Actual Placement and Treatment Details for MPF02

Actual Placement	
Proppant mass placed (2 wing)	136,965
Proppant number, N_{prop}	0.2180
Dimensionless PI, J_{Dact}	0.57
Dimensionless fracture cond, C_{fD}	1.5
Half length, x_f, ft	248.0
Propped width, w, inch	0.13
Post treatment pseudo-skin factor, s_f	–5.49
Folds of increase of PI	4.12
Treatment Details	
Efficiency, eta, %	36.1
Pumping time, t_e, min	58.0
Pad pumping time, t_e, min	27.2
Exponent of added proppant concentration, e_{ps}	0.4694
Uniform proppant concentration in frac at end, lb_m/ft^3	58.2
Areal proppant concentration after closure, lb_m/ft^2	1.1
Max added proppant concentration, lb per gal clean fluid	12.0
Net pressure at end of pumping, psi	122.9

Increasing the maximum possible proppant concentration did the trick. It is now possible to place the required quantity of proppant (i.e., the higher concentration allows us to place more proppant in the same width). In fact, this does not even require all of the capabilities of the equipment; a 14 ppga maximum proppant concentration would be enough.

The actual design now meets the theoretical optimum shown in Table 8-10. The question as to whether or not the larger treatment is justified, however, is still open. Careful economic calculations are needed to rationalize the larger treatment, which would cost about 50 percent more and yield a post treatment skin of –5.54 instead of the –5.50 calculated in the base case. Because the difference is clearly in the error margin, it is difficult to believe that a manager would decide on the more expensive (and more risky) larger treatment.

TABLE 8-12.   Actual Placement for MPF02 (max possible conc: 16 ppga)

Actual Placement	
Proppant mass placed (2 wing)	150,000
Proppant number, N_{prop}	0.2387
Dimensionless PI, J_{Dact}	0.58
Dimensionless fracture cond, C_{fD}	1.7
Half length, x_f, ft	248.0
Propped width, w, inch	0.14
Post treatment pseudo-skin factor, s_f	−5.54
Folds of increase of PI	4.23
Treatment Details	
Efficiency, e_{ta}, %	64.0
Pumping time, t_e, min	32.7
Pad pumping time, te, min	7.2
Exponent of added proppant concentration, e_{ps}	0.2191
Uniform proppant concentration in frac at end, lb_m/ft^3	63.7
Areal proppant concentration after closure, lb_m/ft^2	1.2
Max added proppant concentration, lb per gal clean fluid	13.9
Net pressure at end of pumping, psi	122.9

Proppant Embedment: MPF03

It is widely accepted that in softer formations a considerable part of the injected proppant might be "lost" because it is embedded into the formation wall. Some estimates of width because of embedment are as high as 30 percent (Lacy, 1994).

Let us assume that the rock mechanics lab measured a 33.3 percent embedment for the given formation and closure stress. How can we incorporate this into the design?

The easiest way is to say that the proppant pack permeability (now 60,000 md) will apparently be reduced to 40,000 md. Changing just one line of input from our final base case design (h_f = 211 ft) yields the results shown in Table 8-13:

Proppant pack permeability, md	40,000

TABLE 8-13.   Theoretical Optimum for MPF03

Optimum Placement without Constraints	
Proppant number, N_{prop}	0.1280
Dimensionless PI, J_{Dopt}	0.50
Optimal dimensionless fracture cond, C_{fDopt}	1.6
Optimal half length, x_{fopt}, ft	185.2
Optimal propped width, w_{opt}, inch	0.2
Post treatment pseudo-skin factor, s_f	−5.23
Folds of increase of PI	3.60

Now the maximum possible dimensionless productivity index is less, only 0.50, but even this cannot be realized as the error message indicates (see Table 8-14):

Suboptimal placement with constraints satisfied
Mass of proppant reduced

In fact, only 65,300 lb proppant can be placed because the width at 185 ft is less than it was at 211 ft, and because we need more width to compensate for the loss of conductivity (caused by embedment).

TABLE 8-14.   Actual Placement for MPF03

Actual Placement	
Proppant mass placed (2 wing)	65,285
Proppant number, N_{prop}	0.0929
Dimensionless PI, J_{Dact}	0.46
Dimensionless fracture cond, C_{fD}	1.2
Half length, x_f, ft	185.2
Propped width, w, inch	0.11
Post treatment pseudo-skin factor, s_f	−5.06
Folds of increase of PI	3.31

To make the design possible, we depart from the optimum, multiplying the theoretical optimum length by a factor. Say, for example, that we wish to target 250 ft of fracture length. Still using the 2:1 aspect ratio as most probable, we change the fracture height to 250 ft and then find a factor that results in a half length of 250 ft. The value is 1.58 (see Tables 8-15 and 8-16).

The message shows that the suboptimal placement with the forced modification can be realized (Table 8-17):

Suboptimal placement with constraints satisfied

Length modified

TABLE 8-15. New Height Input and Constraints for MPF03

Fracture height, ft	250
. . .	
Max possible added proppant concentration, lb_m/gal neat fluid	12
Multiply opt length by factor	1.58
Multiply Nolte pad by factor	1

TABLE 8-16. Output for MPF03 (h_f = 250 ft)

Output	
Optimum Placement without Constraints	
Proppant number, N_{prop}	0.0947
Dimensionless PI, J_{Dopt}	0.46
Optimal dimensionless fracture cond, C_{fDopt}	1.6
Optimal half length, x_{fopt}, ft	158.9
Optimal propped width, w_{opt}, inch	0.1
Post treatment pseudo-skin factor, s_f	−5.08
Folds of increase of PI	3.34

TABLE 8-17. Additional Output for MPF03 (h_f = 250 ft)

Actual Placement	
Proppant mass placed (2 wing)	90,000
Proppant number, N_{prop}	0.0947
Dimensionless PI, J_{Dact}	0.44
Dimensionless fracture cond, C_{fD}	0.7
Half length, x_f, ft	251.0
Propped width, w, inch	0.08
Post treatment pseudo-skin factor, s_f	–4.98
Folds of increase of PI	3.19

Under this scenario, we can place the entire 90,000 lb$_m$ of proppant, but the "success" is questionable. With all of the proppant placed, we still create only a –4.98 equivalent skin; the 65,300 lb placed without a forced length constraint actually results in a more attractive –5.06 skin value.

In the presence of significant non-Darcy effects, the proppant number of N_{prop} = 0.095 should be reduced to N_{prop} = 0.05 or, in extreme cases, N_{prop} = 0.01. If we want to compensate for the loss of productivity, we have to *increase* the amount of proppant placed into the pay by the same factor.

By now the reader might feel why we call our approach *Unified Fracture Design*. Systematic use of the proppant number and the optimality criterion makes design trade-offs quite simple and transparent.

Fracture Design for High Permeability Formation: HPF01

In high permeability formations, the optimality criterion will result in a short and wide fracture. To have a basis for comparison, we will use the previous data set except for the following variables: permeability, plane strain modulus, spurt loss, and leakoff coefficient. The input data are summarized in Table 8-18.

The fracture height entry is still left empty. We know that the gross pay interval is 100 ft. A reasonable assumption for high permeability

TABLE 8-18. Input Data for HPF01

Proppant mass for (two wings), lb_m	90,000
Sp grav of proppant material (water = 1)	2.65
Porosity of proppant pack	0.38
Proppant pack permeability, md	60,000
Max propp diameter, D_{pmax}, inch	0.031
Formation permeability, md	50
Permeable (leakoff) thickness, ft	76
Well radius, ft	0.25
Well drainage radius, ft	745
Pre-treatment skin factor	0.0
Fracture height, ft	
Plane strain modulus, E' (psi)	7.5E + 05
Slurry injection rate (two wings, liq + prop), bpm	20.0
Rheology, K' (lb_f/ft^2) × $s^{n'}$	0.07
Rheology, n'	0.45
Leakoff coefficient in permeable layer, ft/min$^{1/2}$	0.01
Spurt loss coefficient, S_p, gal/ft^2	0.02

fracturing, in the absence of any reliable data on stress contrast, is that extensive height growth will not occur if the target fracture length is less than half of the fracture height. At this point, we put a starting estimate of h_f = 100 ft in our design spreadsheet, and specify the operational constraints and parameters shown in Table 8-19. The results are shown in Table 8-20.

TABLE 8-19. Additional Input For HPF01

Max possible added proppant concentration, lb_m/gallon fluid	16
Multiply opt length by factor	1
Multiply pad by factor	1

TABLE 8-20. Theoretical Optimum for HPF01

Optimum Placement without Constraints	
Proppant number, N_{prop}	0.0121
Dimensionless PI, J_{Dopt}	0.31
Optimal dimensionless fracture cond, C_{fDopt}	1.6
Optimal half length, x_{fopt}, ft	56.7
Optimal propped width, w_{opt}, inch	0.9
Post treatment pseudo-skin factor, s_f	–4.05
Folds of increase of PI	2.27

The first design attempt yields a proppant number of 0.012. This is a typical situation for high permeability formations; even with a considerable amount of proppant and well contained fracture height, the proppant numbers will not be very high (Table 8-21). The message says that:

> **Suboptimal placement with constraints satisfied**
>
> Mass of proppant reduced

In fact only 10,700 lb_m of proppant can be placed into the formation if the target length is 56.7 ft. Such a treatment would yield an

TABLE 8-21. Actual Placement for HPF01

Actual Placement	
Proppant mass placed (2 wing)	10,702
Proppant number, N_{prop}	0.0014
Dimensionless PI, J_{Dact}	0.21
Dimensionless fracture cond, C_{fD}	0.2
Half length, x_f, ft	56.7
Propped width, w, inch	0.11
Post treatment pseudo-skin factor, s_f	–2.50
Folds of increase of PI	1.53

extremely low proppant number and an equivalent skin of –2.5. In most cases, this would not be satisfactory, especially because other factors can further decrease the stimulation effect.

Even though this is a relatively soft formation, the fracture width (or lack of width) created during normal propagation severely limits the volume of proppant that can be placed. Note that we have already used a maximum proppant concentration of 16 ppga, but that is not enough.

The solution to the problem is to design a TSO treatment. Knowing that the formation is soft and relatively unconsolidated, we can intentionally arrest fracture propagation at the target length (56.7 ft) and inflate the fracture from there on.

For the TSO design, we re-use the previous input, with just one additional parameter:

TSO criterion Wdry/Wwet	0.7

This TSO criterion suggests that we anticipate the fracture to stop propagating when the "wet width" because of fluid loss (in other words, dehydration) is sufficiently close to the "dry width." The dry width is defined as the width of the fracture after all fluid has leaked off, while the wet width is the width during the treatment when part of the proppant-carrying fluid still has not leaked off. We typically use 0.7 as the critical ratio, but depending on the actual fracture shape and proppant type, the value might differ (Table 8-22).

TABLE 8-22. Actual Placement for HPF01 (with TSO design)

Actual Placement	
Proppant mass placed (2 wing)	90,000
Proppant number, N_{prop}	0.0121
Dimensionless PI, J_{Dact}	0.3127
Dimensionless fracture cond, C_{fD}	1.64
Half length, x_f, ft	56.7
Propped width, w, inch	0.9282
Post treatment pseudo-skin factor, s_f	–4.05
Folds of increase of PI	2.27

The new output suggests that we could place all of the proppant in a 57 ft fracture using a TSO design. This is achieved (internally) by adjusting the proppant schedule so that the proppant concentration in the fracture reaches its critical value at the same time the (unrestricted) lateral extension reaches the target length (Table 8-23).

In fact, 11,000 lb_m of proppant is placed in the fracture in less than 8 minutes. After that, the fracture length remains constant and only the width is increased (Figure 8-3).

The net pressure is considerable, almost 500 psi at the end of the treatment. This is anticipated because the optimum placement calls for an almost 1-inch propped fracture width.

Extreme High Permeability: HPF02

Permeabilities of several hundred millidarcies are not uncommon in naturally fractured formations. To investigate this territory, we repeat the previous design with just one new input (Table 8-24):

Formation permeability, md	500

TABLE 8-23.   Actual Placement for HPF01 (with TSO Design)

Treatment Details	
Pad pumping time, min	0.41
TSO time, min	7.9
Total pumping time, min	24.8
Mass of proppant in frac at TSO, lb_m	11,065
Added proppant concentration at TSO, c_a, lb_m/gal liq	2.0
Half length at TSO, x_f, ft	56.7
Average width at TSO, inch	1.2
Net pressure at TSO, psi	81.1
Max added proppant concentration at end, lb_m/gal-liq	16.0
Areal proppant concentration after closure, lb_m/ft^2	1.3
Net pressure at end of pumping, psi	482

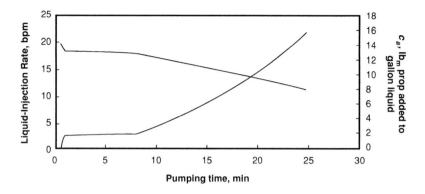

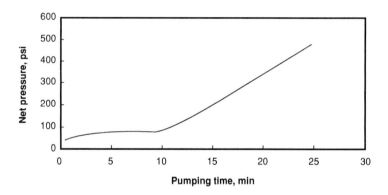

FIGURE 8-3. Fluid, proppant schedule, and net pressure forecast for the TSO treatment.

TABLE 8-24. Theoretical Optimum for HPF02

Optimum Placement without Constraints	
Proppant number, N_{prop}	0.0012
Dimensionless PI, J_{Dopt}	0.23
Optimal dimensionless fracture cond, C_{fDopt}	1.6
Optimal half length, x_{fopt}, ft	17.9
Optimal propped width, w_{opt}, inch	2.9
Post treatment pseudo-skin factor, s_f	−2.90
Folds of increase of PI	1.67

As we see, the target length is now 18 ft. While the program can generate a TSO design for this extreme short fracture length, it would result in an unacceptably high net pressure (cf. the last line of Table 8-25).

Several parameters actually have unrealistic values in the results of the first attempt. The extremely short fracture, even if it could be realized, would not be necessarily useful; the near-wellbore damage might be still dominating at such distances. A reasonable design would call for longer fracture. From an operational viewpoint, net pressure limitations are the most important constraint in high permeability fracturing. A maximum allowable net pressure should be specified to ensure safe operations. A typical value would be 1,000 psi. Let's modify our design in order to satisfy this limitation.

TABLE 8-25.   First Attempt for HPF02

Actual Placement	
Proppant mass placed (2 wing)	90,000
Proppant number, N_{prop}	0.0012
Dimensionless PI, J_{Dact}	0.2299
Dimensionless fracture cond, C_{fD}	1.64
Half length, x_f, ft	17.9
Propped width, w, inch	2.9351
Post treatment pseudo-skin factor, s_f	–2.90
Folds of increase of PI	1.67

Treatment Details	
Pad pumping time, min	0.06
TSO time, min	1.2
Total pumping time, min	18.6
Mass of proppant in frac at TSO, lb_m	2,353
Added proppant concentration at TSO, c_a, lb_m/gal liq	3.0
Half length at TSO, x_f, ft	17.9
Average width at TSO, inch	5.4
Net pressure at TSO, psi	54.5
Max added proppant concentration at end, lb_m/gal-liq	16.0
Areal proppant concentration after closure, lb_m/ft^2	0.9
Net pressure at end of pumping, psi	2142

We have several options. One possibility is to depart from the optimum length, that is, multiplying it by a factor. A realistic design would try to keep the 1:1 aspect ratio; therefore, we select:

| Multiply opt length by factor | 3 |

This would result in an actual placement as shown in Table 8-26. Such a treatment already satisfies the net pressure constraint.

The calculated design calls for starting the addition of proppant almost from the beginning of the treatment. Unfortunately, the design depends heavily on the selected TSO criterion and on the accuracy of the leakoff description. In reality, it is difficult to predict the TSO with such accuracy. It is an art to deliberately arrest fracture propagation while avoiding a near-wellbore screenout (that would prematurely end the treatment). This often requires intuition and experience on the part

TABLE 8-26.   HPF02 with Modified Length

Actual Placement	
Proppant mass placed (2 wing)	90,000
Proppant number, N_{prop}	0.0012
Dimensionless PI, J_{Dact}	0.2058
Dimensionless fracture cond, C_{fD}	0.18
Half length, x_f, ft	53.8
Propped width, w, inch	0.9784
Post treatment pseudo-skin factor, s_f	−2.39
Folds of increase of PI	1.49
Treatment Details	
Pad pumping time, min	0.38
TSO time, min	7.2
Total pumping time, min	24.2
Mass of proppant in frac at TSO, lb_m	10,308
Added proppant concentration at TSO, c_a, lb_m/gal liq	2.1
Half length at TSO, x_f, ft	53.8
Average width at TSO, inch	1.3
Net pressure at TSO, psi	79.7
Max added proppant concentration at end, lb_m/gal-liq	16.0
Areal proppant concentration after closure, lb_m/ft^2	1.3
Net pressure at end of pumping, psi	521

of the fracturing engineer. In the current case, there is another possibility that would minimize the risks associated with the treatment: reduce the amount of proppant and multiply the optimum length by a factor, at the same time (Tables 8-27 and 8-28).

There is little to lose when we reduce the proppant number from 0.0012 to 0.0006. In this range of proppant numbers, the dimensionless

TABLE 8-27.   Testing a Different Set of Inputs for HPF02

Proppant mass for (two wings), lb_m	45,000
. . .	
Multiply opt length by factor	4

TABLE 8-28.   HPF02 with Less Proppant and Modified Length

Proppant number, N_{prop}	0.0006
Actual Placement	
Proppant mass placed (2 wing)	45,000
Proppant number, N_{prop}	0.0006
Dimensionless PI, J_{Dact}	0.1847
Dimensionless fracture cond, C_{fD}	0.10
Half length, x_f, ft	50.7
Propped width, w, inch	0.5189
Post treatment pseudo-skin factor, s_f	−1.84
Folds of increase of PI	1.34
Treatment Details	
Pad pumping time, min	0.34
TSO time, min	6.5
Total pumping time, min	14.0
Mass of proppant in frac at TSO, lb_m	9,523
Added proppant concentration at TSO, c_a, lb_m/gal liq	2.1
Half length at TSO, x_f, ft	50.7
Average width at TSO, inch	0.6
Net pressure at TSO, psi	78.1
Max added proppant concentration at end, lb_m/gal-liq	16.0
Areal proppant concentration after closure, lb_m/ft^2	1.2
Net pressure at end of pumping, psi	239

productivity index is relatively insensitive to proppant volumes or departure from the optimum length, as a matter of fact. Only a moderately negative equivalent skin factor can be realized at such low proppant numbers. This explains the widely accepted view in extreme high permeability fracturing that the most important issue is to "get behind the damage" and create a "halo" (proppant pack) around the screen. The actual fracture length has less significance. Many high permeability fracturing treatments use only 50,000 lb_m or less of proppant.

Low Permeability Fracturing: LPF01

To maintain consistency with our previous examples, we consider a low permeability formation with most of the input parameters similar to our base case as shown in Table 8-29.

TABLE 8-29. Input for LPF01

Proppant mass for (two wings), lb_m	90,000
Sp grav of proppant material (water = 1)	2.65
Porosity of proppant pack	0.38
Proppant pack permeability, md	60,000
Max propp diameter, D_{pmax}, inch	0.031
Formation permeability, md	0.5
Permeable (leakoff) thickness, ft	76
Well radius, ft	0.25
Well drainage radius, ft	745
Pre-treatment skin factor	0.0
Fracture height, ft	
Plane strain modulus, E' (psi)	2.00E + 06
Slurry injection rate (two wings, liq + prop), bpm	20.0
Rheology, K' (lb_f/ft^2) $\times s^{n'}$	0.0700
Rheology, n'	0.45
Leakoff coefficient in permeable layer, $ft/min^{0.5}$	0.00200
Spurt loss coefficient, Sp, gal/ft^2	0.00100
Max possible added proppant concentration, lb_m/gal neat fluid	12
Multiply opt length by factor	1
Multiply Nolte pad by factor	1

Again we will start the design by specifying a fracture height of 100 ft (Table 8-30).

The proppant number is large because of the large contrast in permeabilities. At such a large proppant number, the indicated fracture half length is nearly as long as the side length of the drainage area— explaining why the optimum dimensionless fracture conductivity is significantly larger than 1.6.

If such a fracture could be realized, an extremely large dimensionless productivity index would be established. Unfortunately, there is little chance that a fracture with aspect ratio 8:1 could be created without height increase. In practice, an aspect ratio of about 2:1 is much more likely to be created.

Therefore, we base our design on an assumed aspect ratio of 2:1. Changing the fracture height to 300 ft, the theoretical optimum values become more realistic; the decrease of volumetric proppant efficiency reduces the proppant number (Tables 8-31 and 8-32).

TABLE 8-30.   Theoretical Optimum for LPF01 ($h_f = 100$ ft)

Optimum Placement without Constraints	
Proppant number, N_{prop}	1.2077
Dimensionless PI, J_{Dopt}	1.06
Optimal dimensionless fracture cond, C_{fDopt}	2.9
Optimal half length, x_{fopt}, ft	423.0
Optimal propped width, w_{opt}, inch	0.1
Post treatment pseudo-skin factor, s_f	−6.30
Folds of increase of PI	7.66

TABLE 8-31.   Theoretical Optimum for LPF01 ($h_f = 300$ ft)

Optimum Placement without Constraints	
Proppant number, N_{prop}	0.4026
Dimensionless PI, J_{Dopt}	0.68
Optimal dimensionless fracture cond, C_{fDopt}	1.8
Optimal half length, x_{fopt}, ft	309.4
Optimal propped width, w_{opt}, inch	0.1
Post treatment pseudo-skin factor, s_f	−5.78
Folds of increase of PI	4.92

TABLE 8-32. Actual Placement for LPF01 (h_f = 300 ft)

Actual Placement	
Proppant mass placed (2 wing)	90,000
Proppant number, N_{prop}	0.4026
Dimensionless PI, J_{Dact}	0.68
Dimensionless fracture cond, C_{fD}	1.8
Half length, x_f, ft	309.4
Propped width, w, inch	0.06
Post treatment pseudo-skin factor, s_f	−5.78
Folds of increase of PI	4.92
Treatment Details	
Efficiency, e_{ta}, %	67.1
Pumping time, t_e, min	52.7
Pad pumping time, t_e, min	10.4
Exponent of added proppant concentration, e_{ps}	0.1966
Uniform proppant concentration in frac at end, lb_m/ft^3	22.6
Areal proppant concentration after closure, lb_m/ft^2	0.5
Max added proppant concentration, lb per gal clean fluid	3.5
Net pressure at end of pumping, psi	113.7

While the design is now more realistic, one variable deserves special attention: the fluid efficiency, which increased to 67 percent. Why did this happen? According to our definition, leakoff occurs only in the pay layer (with a net thickness 76 ft). Now that the actual fracture height is taken as 300 ft, only one quarter of the total surface contributes to leakoff, and the efficiency is very high. In reality, this is not likely because the pay interval is bound by continuously and perfectly non-permeable layers. Thus, it is wise to reconsider the leakoff (and spurt loss) parameters once a significant change in fracture height has been introduced.

Repeating the design with adjusted leakoff and spurt loss coefficients,

Leakoff coefficient in permeable layer, ft/min$^{0.5}$	0.0050
Spurt loss coefficient, S_p, gal/ft^2	0.00250

we obtain the results in Table 8-33.

TABLE 8-33.   Actual Placement for LPF01 (h_f = 300 ft, Adjusted Leakoff)

Actual Placement	
Proppant mass placed (2 wing)	90,000
Proppant number, N_{prop}	0.4026
Dimensionless PI, J_{Dact}	0.68
Dimensionless fracture cond, C_{fD}	1.8
Half length, x_f, ft	309.4
Propped width, w, inch	0.06
Post treatment pseudo-skin factor, s_f	−5.78
Folds of increase of PI	4.92
Treatment Details	
Efficiency, e_{ta}, %	38.2
Pumping time, t_e, min	92.8
Pad pumping time, t_e, min	41.5
Exponent of added proppant concentration, e_{ps}	0.4475
Uniform proppant concentration in frac at end, lb_m/ft^3	22.6
Areal proppant concentration after closure, lb_m/ft^2	0.5
Max added proppant concentration, lb per gal clean fluid	3.5
Net pressure at end of pumping, psi	113.7

The fluid efficiency is now more realistic, but the final fracture length and propped width are exactly the same as before. How is it possible that such a large change in the leakoff parameters does not affect the final results? The answer to this question reveals the main difference between *simulation* and *design*. In our design procedure, the target length and target propped width are derived from the reservoir and proppant properties. The leakoff parameters (and other variables) determine how we achieve our final goal, but the goal is the same, whether there is intensive leakoff or not. The change in the leakoff parameters shows up in the actual proppant schedule. Now we have to pump for a considerably longer time.

Experienced fracturing engineers would probably not yet accept the design. The indicated propped fracture width is only 0.06 inch (i.e., less than 3 grain diameters of the 20/40 mesh proppant). A good design ensures a certain minimum width (or a certain minimum areal proppant concentration).

At this point, we must either increase the amount of proppant or depart from the indicated optimum length, now multiplying it by a factor less than one. The advantage of creating a shorter fracture is also evident in the volumetric proppant efficiency; in other words, reducing the aspect ratio to less than 2:1, less proppant will "avoid" the pay. The relevant lines of the input and results are shown in Tables 8-34 and 8-35.

TABLE 8-34. Input for LPF01 (Final Design)

Proppant mass for (two wings), lb_m	90,000
...	
Fracture height, ft	200.0
...	
Leakoff coefficient in permeable layer, ft/min$^{0.5}$	0.0050
Spurt loss coefficient, S_p, gal/ft^2	0.0025
...	
Max possible added proppant concentration, lb_m/gal neat fluid	12
Multiply opt length by factor	0.55
Multiply Nolte pad by factor	1

TABLE 8-35. Actual Placement for LPF01 (Final Design)

Actual Placement	
Proppant mass placed (2 wing)	90,000
Proppant number, N_{prop}	0.6039
Dimensionless PI, J_{Dact}	0.67
Dimensionless fracture cond, C_{fD}	6.7
Half length, x_f, ft	198.3
Propped width, w, inch	0.13
Post treatment pseudo-skin factor, s_f	−5.76
Folds of increase of PI	4.85
Treatment Details	
Efficiency, e_{ta}, %	38.3
Pumping time, t_e, min	38.5
Pad pumping time, t_e, min	17.2
Exponent of added proppant concentration, e_{ps}	0.4457
Uniform proppant concentration in frac at end, lb_m/ft^3	54.3
Areal proppant concentration after closure, lb_m/ft^2	1.1
Max added proppant concentration, lb per gal clean fluid	10.8
Net pressure at end of pumping, psi	166.4

Note that targeting the smaller fracture allowed us to reduce the assumed height as well. Therefore, the design can utilize more efficiently the 90,000 lb$_m$ of proppant. The post-treatment dimensionless productivity index and equivalent skin factor are basically the same as in the previous case. This final design is more practical and certainly easier to carry out.

SUMMARY

In this chapter, we showed many examples of practical fracture design. The concept of proppant number and dimensionless productivity index helped us to make important decisions without going into unnecessary details. The included design spreadsheet was used extensively to consider "what-if" scenarios and investigate options.

In hydraulic fracture design, where the reliability of the available input data is always limited and the process itself is inherently stochastic, it is extremely important to proceed in an evolutionary manner, continuously improving the design. The simple spreadsheet does not substitute for sophisticated 3D fracture simulators, but it certainly provides a flexible tool to make the basic decisions before the final design.

Quality Control
and Execution

Quality control has been embedded in the vernacular of fracturing operations for decades. The motivation for quality control was often poor execution.

Today, the service companies have formalized and extended an entire assortment of self-policing quality control schemes. The term itself, *quality control,* has become rather generic, now used to refer to anything from a checklist, filled-out in the field, to a subtle marketing overture intended to attract clients, to the latest avante garde business psychology.

More to the point, quality control means that a fracture treatment should, and *can,* be carried out as it was designed. This means careful pre-treatment planning; well maintained and functioning equipment; trained, conscientious and well-informed personnel; intense tracking of each fracturing material and critical treatment parameters; and post-treatment evaluation.

And the results speak for themselves: service company performance on fracture treatments, while certainly not perfect, is as good as it has ever been.

Note: A. S. Demarchos contributed to this chapter.

FRACTURING EQUIPMENT

While often overlooked, the fracturing equipment is the starting point for successful quality control and execution.

Stimulation equipment has undergone extensive changes since the first commercial hydraulic fracturing treatment was performed in 1949. That job involved hand-mixing five sacks of 20 mesh sand into 20 bbl of fluid (0.6 lb_m/gal proppant concentration). The mixture was pumped downhole with a 300 horsepower triplex pump used in cementing and acidizing.

While treatments have grown in magnitude and complexity—a modern massive hydraulic fracture treatment may involve 10,000 sacks of sand and proppant concentrations of 10 lb_m/gal or higher (Figure 9-1)—the basic configuration has not changed since that first treatment. Proppant and a treating fluid are delivered to a blender where they are mixed and transferred to the high pressure pumps. The proppant-laden treating fluid is then pumped through a high pressure manifold to the well.

The equipment required to perform stimulation treatments are blending, proppant handling, pumping and monitoring/control equipment.

Blending equipment is used to prepare the treatment fluid, combining specified proportions of liquid and dry chemical additives into

FIGURE 9-1. Modern massive hydraulic fracture treatment.

the stimulation fluid. Fracturing fluids are either batch-mixed before the treatment (and held in frac tanks until it is needed) or mixed continuously throughout the treatment. For continuous mixing, the base fluid is prepared by a pre-blender, which combines a liquid gel concentrate with mix water and provides sufficient hydration time to yield the required base fluid gel viscosity. The hydrated gel is then pumped from the hydration tank to the blender where the additives and proppant are combined with the treatment fluid.

The quality of the mixing process is now almost always computer controlled. Set points for the mixture concentrations are entered into a computer and maintained automatically, regardless of the mixing flow rate. Operational parameters of the blender such as tub level, mixer agitation, and pressures also have been placed under automatic control, thereby minimizing the potential for human error.

Proppants are stored on location, transferred and delivered to the blender using several methods. Sacked proppant can be handled manually, or delivered using dump trucks/trailers and pneumatic systems. Ever increasing quantities of proppant have necessitated the use of field storage bins. Eventually, when the proppant volume exceeds the capacity of a single field storage bin, multiple bins are located around a gathering conveyor that transfers proppant to the blender. Given the distances that proppant may travel from the farthest storage bin to the blender, automatic control systems have been added to the storage bins and gathering conveyors to allow uninterrupted delivery of proppant.

The nominal 300-horsepower pump used in 1949 has been replaced today by pumps with 2,000-plus horsepower produced from a single crankshaft. The pressure requirements for treatments also have grown from 2,000 psi to, in some extraordinary cases, over 20,000 psi. Transmissions today can be shifted under full power. Computers synchronize the engine speed with the gear shift such that the pump rate before and after the shift are the same. Computer-controlled pumping equipment also allows automatic pressure and/or rate control.

The monitoring of stimulation treatments has also progressed— from the pressure gauges, stopwatches and chart recorders of decades past, to full computer monitoring and control today. Today, more than a thousand individual parameters can be simultaneously monitored and recorded during a stimulation treatment. Monitoring the treatment fluids is an essential element of quality control. Parameters monitored and recorded during a stimulation treatment include but are not limited to pressures, temperatures, flow rates, proppant and additive concentrations, pH, and viscosity. Any or all of these parameters can be displayed during the job, along with, in many cases, real-time translation of the values to downhole conditions. Many equipment parameters—run times, pressures, vibration, and so on—are also monitored and recorded during the treatment. This information is used to diagnose and preempt equipment problems, to assist in equipment maintenance and to improve future equipment layout and designs.

EQUIPMENT LIST

Assembling an appropriate combination of equipment—pumps, blenders, trucks, monitoring and electronic equipment—is vital to the success of any fracture stimulation treatment. Following is a basic list of the fracturing equipment, provided in order from the water source to the wellhead.

Water Transfer and Storage

Pit Manifold

The pit manifold provides suction from the water pit (if used) and a common suction header with at least eight 4-inch suction connections. It is used only when the blender takes its water supply directly from a water pit (cf. Figure 9-8), or when a transfer pump is used to keep fracturing tanks filled.

Water Transfer Pump

Low pressure, high rate pumps are used to transfer fluids from the water pit (or other water supply) to the fracturing tanks and/or blender. Transfer pumps may or may not be needed, depending on the distance between the water source and blender, and their respective elevations. Depending on the size of the treatment and distance from the water source, one or more pumps will be used to transfer water through standard 6-inch PVC irrigation pipe.

Fracturing Tanks

Self-contained 500 barrel tanks are used to store fracturing fluid on location. These "frac tanks" have integral wheels and can be easily mobilized to and between locations. These have a minimum of four 4-inch connections and a 12-inch butterfly valve that is used to header tanks together to provide a common water source. The required number of frac tanks is determined by the size of the treatment. If a combination of a water pit and tanks is to be used, a minimum of four tanks is typical.

Proppant Supply

Sand Supply System (stationary)

A trailer-mounted sand storage unit equipped with a conveyor belt is used to supply the blender with proppant. It is gravity fed and has hydraulically controlled gates. Each unit has at least two separate compartments and is capable of delivering sand from either or both compartments at the same time. Depending on the volume and maximum proppant concentrations employed, more than one unit is sometimes needed; in this case, a central conveyor unit is used to coordinate sand delivery to the blender.

Sand Supply System (mobile)

Truck-mounted sand transport and supply units are typically used for smaller treatments. These units have capabilities similar to the stationary units, except that they hold much less volume, from 35,000 to 60,000 pounds of proppant instead of 250,000 to 500,000 pounds. In the case of very large treatments, a mobile supply unit may be used to feed a much larger stationary system, such as the so-called "mountain mover."

Sand Conveyor

Whether a stationary or mobile sand supply system is used, the layout of the location or the size of the fracturing treatment may dictate the use of a conveyor belt system to transfer proppant to the blender. Most conveyor systems are trailer mounted and can be easily maneuvered into position on location.

Slurification and Blending

Chemical Mixing/Hydration Unit

There are two methods for mixing fluids prior to pumping. First, they can be batch mixed in frac tanks. This allows the quality and consistency of the fluids to be easily controlled, which many operators appreciate, but there are disadvantages. If there are any delays in pumping, gelled fluids can deteriorate rapidly, especially in warm ambient temperatures. Also, unused fluids introduce certain environmental considerations and must be disposed of properly.

The second method is to mix fluids as needed, "on the fly." Additives, gelling, and cross-linking agents are blended together in a *hydration tank* with water to form the fracturing fluid. The *hydration unit* must be carefully operated so the gel has adequate residence time and is fully hydrated before it is delivered to the blender. (One of the problems of mixing without modern hydration units is that the pH must be adjusted for proper hydration. This is a very delicate operation, especially at high ambient temperature.) A properly maintained hydration unit can effectively mix both dry and liquid chemical additives. Use of a *continuous hydration unit* minimizes problems associated with the mixing of polymers, which are added in liquid-slurry form—thus eliminating the need to use the blender for agitation. Associated metering pumps, hooked up to the main injection lines, are needed to introduce additives to the fracturing fluid. Operating data from the hydration unit is transmitted by cable to the control center and monitored continuously during the treatment.

Blender

A self-contained, truck-mounted blender combines the water, gel, sand, and other additives into one uniform mixture. The blender sits at the heart of the fracturing treatment (Figure 9-2). It is connected to the fracturing fluid supply with at least four and as many as twelve 4-inch flex hoses. The discharge side is connected to the low pressure inlet of the fracturing manifold with 4-inch flex hoses, or directly to the frac pumps if the treatment is small. The blender must be capable

FIGURE 9-2. The blender sits at the "heart" of the fracturing treatment.

and calibrated to add dry and liquid additives at very precise rates. Blender performance is defined by the volume and rate at which it can accept proppant. A dual blender configuration can be employed in treatments that demand high proppant concentrations and high rates. Data from the blender is also transmitted via cable to the control center.

Pumping

HI-LO Pressure Manifold

The HI-LO pressure manifold can be truck- (Figure 9-3), trailer- or skid-mounted. The low pressure ("LO") header is used to couple the discharge of the blender with the suction side of the frac pumps. Four to eight 4-inch flex hoses run from the blender to the manifold, and additional flex hoses run to the individual frac pump inlets. A standard manifold can service eight frac pumps simultaneously. All connections on the low pressure header are equipped with butterfly isolation valves.

The high pressure side ("HI") of the manifold is fed by high-pressure steel pup joints from the discharge the frac pumps and, in turn, connects to the wellhead with high pressure pump joints. A fracturing valve or tree-saver is used to make the physical connection

FIGURE 9-3. Truck mounted HI-LO pressure manifold.

at the wellhead. Each line to and from the high pressure header employs (in series) a 15,000 psi check valve to control fluid movement and a 15,000 psi plug (isolation) valve.

High Pressure Manifold

For small fracturing treatments (i.e., when the frac pump intakes are connected directly to the blender), a simple high pressure (one-sided) manifold is used to couple the frac pump discharges and the treatment well. Again, high pressure check valves and plug valves are employed.

Frac Pumps

If the blender is the *heart*, then frac pumps provide the *muscle* to a fracturing treatment (Figure 9-4). These pumps take in the fracturing fluid at low pressure (about 60 psi) and discharge it at the required pressure (1,000s of psi). These positive displacement plunger-type pumps are available in several sizes. The triplex configuration (three plungers) is most common. Quintaplex frac pumps (5 plungers) are gaining popularity and, of course, are capable of handling more fluid and at a higher pressure than the triplex. Hydraulic horsepower for

FIGURE 9-4. Frac pumps are the "muscle" of the fracturing treatment.

these pumps ranges from less than 1,000 HHP for an early model triplex to well over 2,000 HHP for a late model quintaplex.

Frac pumps are truck- or trailer-mounted. They are equipped with high pressure shut-downs and should be controllable by wire from a remote position.

High Pressure Steel

High pressure pup joints, hammer unions, Y-unions, swivels (or "chicksans"), check valves, high pressure pop-off valves, and plug valves are required to connect the discharge side of the frac pumps to the manifold and the manifold to the well. These pieces, often collectively referred to as the "treating iron," are available in 2-, 3- and 4-inch diameters and with an assortment of pressure ratings.

A *Y-union* is often used near the wellhead to bring two high pressure lines from the manifold together to a single injection point (cf. Figure 9-6). *Check valves* isolate the fracturing equipment from injection well backpressure. If for any reason pressure in the fracturing line exceeds a maximum set pressure, a *pop-off valve* opens to relieve pressure and prevent equipment damage or personal injury. A *plug valve* is also used in-line and upstream of the wellhead as an added control point. To minimize the effect that pipe vibration and movement has on the rigid connections, all high pressure equipment is connected using a minimum of two pup joint sections with a *chicksan* in the middle. Additional chicksans are often used to simplify hookup and further minimize vibration effects.

Flex Hoses

A 4-inch flex hose, rated at 150 psi and normally operated at 60 psi, is typically used to connect the water source to the blender and the blender to the manifold, and to supply fracturing fluid to the pump intakes; 12-inch flex hoses are normally used to header the frac tanks together to create a common water source.

Monitoring and QA/QC

Frac Van

All of the equipment, flow rates and critical pressures are monitored by a central command post, often referred to in vernacular as the "frac van" (Figure 9-5). Data is displayed, recorded, processed, and printed

FIGURE 9-5. Vital data are continuously monitored in the "frac van."

minute-by-minute in the frac van. The "treater" is an individual responsible to monitor the flow of data from a programmable display and control panel. As a minimum, the display continuously shows slurry rate, proppant concentration, wellhead treating pressure, and elapsed treatment time.

Vans equipped with multiple displays and parallel processing capability allow the simultaneous processing and evaluation of treatment data in real-time (e.g., calculating bottomhole pressures or fluid transit times, or graphically monitoring the evolution of various diagnostic plots during the treatment).

Quality Control Van

A mobile chemical laboratory is utilized to "catch samples" and test them before and during the treatment. A typical mobile laboratory includes, as a minimum: pH meter; temperature probe; proppant sieves and mechanical shakers; laboratory scale; blender; water bath; viscometer and possibly an inline viscometer; and miscellaneous supplies such as cups, stirrers, gloves, filters, reference manuals, and a microwave oven. The van is typically equipped with a generator to power all of the equipment.

Communications

All operators of any machinery, the person monitoring the water supply, and any other personnel critical to the fracturing operation must be in constant two-way communication with the treater, at all times. Communication equipment is typically integrated to the command center, and the manufacturers of these units offer several options.

Remote Monitoring

Remote monitoring brings the well site to the client by providing real-time communication via satellite. Satellite up-link capability is becoming a priority for all geographically distributed fracturing operations. The manufacturers of command centers now offer a satellite option.

Remote Operations

When multiple frac pumps are used (almost always), the number of operators can be limited using remote control boxes. Each control box handles up to four pumps, and one operator can normally operate two boxes. Newer equipment can be wired directly into the command center, although this is not necessary.

Miscellaneous

A *steel carrier* equipped with a truck-mounted crane/hoist is used to transport all the high pressure steel and assist with the assembly.

Operations as complex as fracturing treatments should be planned during daylight. Occasionally, either due to delays, mechanical failures, or the size of the treatment, night operations are required. For this situation, self-contained *frac lights* are required. These are high intensity lights, the style used in sport stadiums, mounted on telescoping poles and each set with its own generator.

Steel stakes and steel ropes are used to secure all high pressure discharge lines during the fracturing treatments. High pressure treating iron that is not properly staked can whip around uncontrollably if it ruptures during pumping, with great potential to damage equipment or injure personnel.

While there is really no typical assembly of fracturing equipment— the assembly or "frac spread" varies widely by geographic location and the anticipated pressures, temperatures, and volumes associated with a given treatment—Table 9-1 provides an example of equipment that might be included in a relatively small spread.

SPECIAL INSTRUCTIONS ON HOOK-UP

Several potential configurations for the layout of fracturing equipment at the well site are provided in Figures 9-6 to 9-8. The hook-up configuration is often dictated by variables such as treatment size, water supply source, surface location, and availability of equipment. The following narrative describes Figure 9-6, which is really the ideal layout, to be used when conditions permit.

Spotting the Equipment

1. Identify a flat and level area large enough to hold all of the necessary frac tanks—at a distance far enough from the well to allow placement of the hydration unit, blenders, frac pumps,

TABLE 9-1. Example of a "Frac Spread"

Equipment	Specifications	Qty	Comments
Frac tanks	500 barrels each	6–8	
Blender	120 barrels per minute (bpm)	1	Or two 60 bpm
Frac pumps	2,000 HHP, 14 bpm	5	One is for stand-by
Manifold	10,000 psi	1	Trailer mounted
Flex hoses	4-inch, 30 ft, 125 psi and 3-inch, 30 ft, 125 psi	28 of each	12 each for pumps, 16 each for blender and hydration
Pup joints	3- or 4-inch, 10,000 psi, 8 ft length	20	12 for the pumps, 2 for the frac line, 6 spares
Pup joints	3- or 4-inch, 10,000 psi, 2, 3, 4 ft length	4 each	
Pup joints	3- or 4-inch, 10,000 psi, 20 ft length	8	4 for frac line, 2 for relief line, 2 spares
Swivels	3- or 4-inch, 10,000 psi	30	3 for each pump, 6 for frac line, 9 spares
Y-unions	3- or 4-inch, 10,000 psi	5	4 spares
Pop-off valves	3- or 4-inch, 10,000 psi	3	2 spares
Plug valves	3- or 4-inch, 10,000 psi	11	8 spares
Check valves	3- or 4-inch, 10,000 psi	5	3 spares
Pump control boxes	4 frac pumps each	3	1 spare
Rate transducers	10,000 psi	3	2 spares
Pressure transducers	10,000 psi	3	2 spares
Command center	Per vendor	1	
Two-way radios	2 mile range	16	One for each critical person on location, 8 spares
Satellite up-link	Per vendor	1	More companies requiring remote monitoring
Respirator gear	Per vendor	4	For person handling toxic chemicals, 2 spares
Safety gear	Hard hats, goggles, gloves	12 sets	One set per person
Frac lights	High power, telescoping	4	

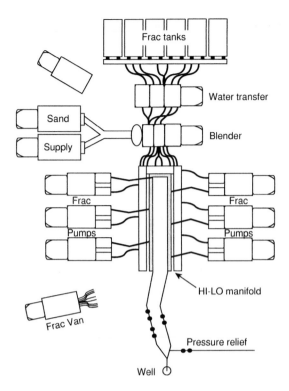

FIGURE 9-6. The most desirable fracture equipment layout, "on the fly" with HI-LO manifold.

HI-LO pressure manifold, and high pressure fracturing line *between* the frac tanks and the wellhead.

2. Draw a straight line from the well to this area. At the head of the line, draw a line perpendicular to it. Center all frac tanks here. It is important that the tanks are level.

3. Place the hydration unit, if one is used, in front of and centered with respect to the frac tanks. If fluids are batch-mixed, place the blender(s) here instead.

4. Place the blender(s) parallel to the hydration unit. If two blenders are used, place blenders side-by-side.

5. Place the sand supply system in line with the hopper on the blender(s), backing it into position.

6. Along the line from the wellhead and near the blender, place the HI/LO pressure manifold.

7. At the discharge side of the manifold, place the high pressure fracturing line.

8. On each side of the HI-LO pressure manifold, back in and position the frac pumps (more details on hook-up are provided below).
9. Place the monitoring van off to one side in a position that offers a full view of the well and pumping equipment.
10. Lay the high pressure pop-off line away from all personnel and equipment.
11. Place the QA/QC van near the hydration unit and blender(s).

Fluid Supply-to-Blender

The frac tanks should be placed on the same level and connected together with 12-inch flex hoses, creating a common manifold and ensuring an uninterrupted supply of fluid. Connect the hydration unit to the manifold with 4-inch flex hoses. Under ideal conditions, a 4-inch flex hose can deliver up to 8 bpm, but the number of hoses required depends on the fluid rate, viscosity, and distance from the source. All flex hoses should be free of any kinks and obstructions. Connect the discharge side of the hydration unit to the suction side of the blender with 4-inch flex hoses. Connect the blender discharge to the low pressure side of the HI-LO manifold with 4-inch flex hoses. Again, use the 8-bpm rule-of-thumb to determine how many hoses are needed.

Proppant Supply

The idea here is to have the conveyor belt from the sand transport feeding the blender's hoppers. On small jobs, a trailer-mounted transport might be backed right up to the blender. When a stationary proppant supply unit is used, it should be placed in such a way that it can be easily filled by trailer-mounted transports as needed.

Frac Pumps

Each pump intake is connected with one 3-inch or 4-inch flex hose to the low pressure (supply) side of the frac manifold. The flex hose must be small enough to maintain fluid velocity and prevent sanding, yet large enough that it does not restrict flow. The discharge of each pump is connected to the high pressure side of the manifold with at least two pup joint sections and a chicksan (swivel) in between. Preferably, a chicksan is used at the pump discharge, then a pup joint, chicksan, pup joint, and a third chicksan at the manifold. This allows enough movement so that the treating iron does not loosen up or rupture as it vibrates and shifts under high pressure.

Manifold-to-Well

Each high pressure discharge outlet of the HI-LO manifold, whether one or two, is connected to the wellhead, again using multiple pup joints and chicksans to avoid any rigid lines (Figure 9-7). A plug valve, check valve, pressure transducer, and rate transducer are placed in each line as close to the wellhead as possible. If two lines are used, they should be joined using a Y-union at the wellhead. The arrow on the check valve should point in the direction of flow (i.e., toward the well) to avoid an all-too-common mistake. If a flapper-type check valve is used, make sure the valve is placed right side up and level. At the wellhead, a tree-saver or a frac valve may be used. The high pressure pop-off valve should be placed before the check valve and set to the maximum pressure the well dictates. A high pressure bleed-off line is

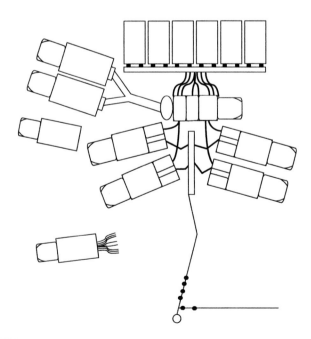

FIGURE 9-7. Fracture equipment layout, batch mix with no HI-LO manifold.

connected to this valve and directed away from the well and equipment. The bleed-off line and the high pressure lines from the manifold to the well should be securely staked.

Monitoring/Control Equipment and Support Personnel

Though most equipment can be connected and monitored from the frac van, it should still be positioned such that the treater has an unobstructed view of all critical components.

As a minimum, the blender, hydration unit, and transducers from the high pressure treating line should be connected to the monitoring van. The frac pumps and the hydration unit can be directly controlled from the van, or by the operators outside. It is normal for one operator to be stationed on the blender and one on the hydration unit. The pumps can be wired to suitcase-style boxes, four at a time, and controlled remotely. This way a single experienced operator can control as many as eight pumps (Figure 9-8).

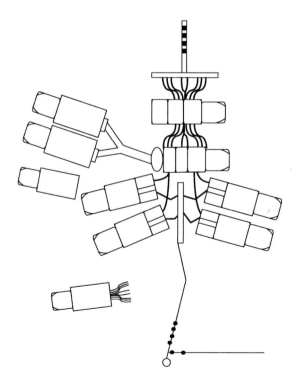

FIGURE 9-8. Fracture equipment layout, "on the fly" with frac pit and no HI-LO manifold.

In the van, the treater coordinates the operators and equipment and generally executes the treatment; the service company engineer oversees any real-time data processing and liaisons with the operating company engineer or representative. Observers can be in the van, if space is available.

Outside, one person is stationed at the wellhead valve, one person at the sand supply system, one at the frac tank manifold, and one on top of the tanks, monitoring fluid levels. If water transfer pumps are used, another person is required to monitor those. The fuel level in all equipment is monitored continuously, usually by the fuel service company representative. The fuel supply truck is placed in a position that allows the operator to refuel equipment as necessary (using a long flex hose).

At least one technician should be in the QA/QC van, capturing samples and monitoring the fracturing fluid quality and proppant concentration. This data is transmitted to the control van. Often, samples are saved for the customer.

All personnel directly involved in the treatment should be equipped with two-way radios.

STANDARD FRACTURING QA PROCEDURES

A number of quality control checks are undertaken prior to each fracture treatment to verify the performance of all fluids and proppants. The treatment itself should also be closely monitored so that (1) to the extent possible, modifications that will improve the outcome of the treatment can be made in real-time and (2) unavoidable deficiencies in the treatment execution can be appropriately evaluated post mortem.

■ *Pre-job Testing*

Prior to pumping, each frac tank is strapped and tested for specific gravity, pH, and temperature. A sample is taken from each tank and tested with gelling agent for viscosity and cross-link time. A composite fluid sample is tested with chemicals from location.

■ *Proppant Validation*

Proppant sieve analysis is provided on location. If proppant does not meet acceptable standards, each compartment is tested individually.

∎ *Pre-job Inventory*

Prior to the start of the job, the Stimulation Treatment Check List is filled out with beginning volumes of all chemicals and frac fluid on location. Proppant storage is visually inspected and compared to weight tickets.

∎ *Job Testing and Recording*

Fluids and chemicals are physically strapped every 5,000 gallons or as often as possible. Samples of the pad and 2–3 slurry stages are taken along with corresponding proppant samples.

∎ *Real-Time QA*

In addition to normal treatment displays of rate, pressure, net pressure, and sand concentration, the following parameters will be displayed and recorded: pH, fluid temperature, viscosity, and all additive rates.

∎ *Post-job Reports*

In addition to the standard treatment outputs, the treatment report includes the following: Proppant Sieve Analysis and QC Form, Water Quality Control Form, Frac Fluid Blending and QC Form, and Stimulation Real Time Report (cf. Appendix F).

Additional quality control and quality assurance measures are provided in Chapter 6 (Fracturing Materials) and Appendix F (Standard Practices and QC Forms).

FORCED CLOSURE

Closing wells in for a matter of hours, overnight, or for several days following a hydraulic fracture treatment was the accepted practice for many years. The extended shut-in time was thought to allow the fracture to close (or "heal"), as well as allowing any viscosified fluids to break completely back to water.

However, fractures, particularly in tight reservoirs, may require a long time to close, and during this time, excessive proppant settling may occur. If the fracture loses conductivity near the wellbore, the treatment may fail. Any pinching effect in the near-wellbore area or decrease in conductivity in the proppant pack may outweigh the time-delay benefit of fluid cleanup in the proppant pack.

For this reason, today a technique called "forced closure" is very often applied. Forced closure consists of flowing fracture fluids back out of the well starting immediately after the end of pumping (within the first minute) at a rate of 10s-of-gallons to several (2 to 3) barrels per minute, depending on the number and size of perforations. Flow rates can be controlled using pressure and choke tables.

Forced closure does not necessarily cause rapid fracture closure (as the name implies), but rather involves something akin to reverse gravel packing of proppant at the perforations. This can be an effective way to prevent proppant settling. While somewhat counterintuitive, experience shows that proppant will not flow back through the perforations even when a well is aggressively flowed back with viscous fluids.

A major benefit of this immediate flowback is that the supercharge of fluid pressure (built up during the fracture treatment) assists in fracture cleanup and establishing production. With the conventional shut-in approach, this pressure dissipates before the well is opened to flow. Forced closure also provides some latitude in the fluid breaker design. Overly aggressive breaker schedules can result in premature loss of fracturing gel viscosity and rapid settling of the proppant. Ideally, a well should initially produce some amount of unbroken gel following a successful fracture treatment.

The dominant mechanism in forced closure is felt to be the creation of a proppant pack opposite the perforations. This would clearly explain the reduced proppant production and improved near-well fracture conductivities that have been observed. Forced closure should also promotes better grain-to-grain contact for treatments that use resin-coated sand.

As a collateral benefit, the artificial pressure built up in the formation by the fracture treatment is often sufficient to clean excess proppant out of the wellbore during the forced closure. This can eliminate costs that would otherwise be incurred in coiled tubing cleanout or sand bailing.

Energized and foam treatments should be flowed back quickly and aggressively to take advantage of the energized gas. Shutting in a fracture treatment that uses CO_2 or N_2 for any period is counterproductive. Reservoirs with any permeability quickly absorb the energizing gas.

QUALITY CONTROL FOR HPF

Many early HPF treatments failed because of equipment problems and a lack of quality control on fluids and proppants. Generally, the intense quality control measures that are standard for onshore massive hydraulic fracture treatments were not immediately adopted on the small offshore *frac & pack* treatments. This invited skepticism of the process and slowed the introduction of HPF technology somewhat. In addition to quality control procedures that were eventually instituted by all major service companies, it became common for producing companies to supply a consultant or in-house specialist to oversee the quality control on most HPF treatments.

10

Treatment Evaluation

REAL-TIME ANALYSIS

Fracturing pressure is often the only direct information available to monitor (or rather, infer) evolution of the fracture during the treatment. Thus, fracture pressure interpretation and decision making are some of the primary responsibilities of the fracturing engineer.

A log-log plot of bottomhole treating pressure versus time suggested by Nolte and Smith (1981) is the classic diagnostic plot used for this purpose (cf. Figure 10-1). After noting the obvious treatment features (e.g., injection rate, fluid quality, proppant concentration), the analysis becomes qualitative.

A steady positive slope of order 0.25 is interpreted as unrestricted (normal) fracture propagation (Type I). A change in the slope from positive to negative denotes an abrupt increase in the fracture surface, as in the case of height growth into another layer (Type II). An increasing slope that approaches unity is considered a sign of restricted tip propagation and is often followed by an even larger slope, indicating the fast fill-up of the fracture with proppant (screen-out).

A rigorous leakoff description and some significant assumptions about fracture geometry are needed to carry out a more quantitative interpretation.

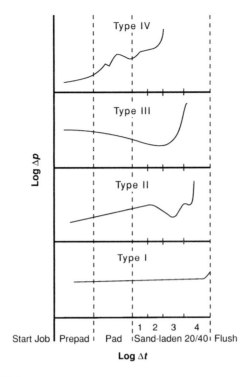

FIGURE 10-1. Real-time pressure response types (Nolte-Smith Plot).

HEIGHT CONTAINMENT

Vertical fracture propagation is constrained by the same mechanical laws as lateral propagation, except that the minimum horizontal stress can vary significantly with depth, and that variation may limit vertical growth.

The equilibrium height concept of Simonson et al. (1978) provides a simple and reasonable method to calculate fracture height when there is a sharp stress contrast between the target layer and the over- and under-burden strata. A minimum horizontal stress that is considerably larger in the adjacent layers (by several hundred psi) tends to contain the fracture height until the critical stress intensity factor is exceeded, either at the top and/or bottom edge of the fracture. As the pressure at the reference point (center of the perforations) increases, the equilibrium penetrations into the upper (Δh_u) and lower (Δh_d) layers increase. The requirement of equilibrium poses two constraints (one at the top, one at the bottom) and the two penetrations can be obtained solving a system of two equations. If the hydrostatic pressure component is neglected, the solution is unique up to a certain pressure called the "run-away pressure." Above the run-away pressure, there is no equilibrium state. This does not suggest unlimited height growth, but rather that the vertical fracture growth is no more constrained than the lateral growth. As a consequence, we can assume that the fracture propagates radially.

In the case of a large *negative* stress contrast (stress in adjacent layers *smaller* than in target layer), unlimited height growth may occur, irreversibly damaging the well.

The equilibrium height concept can be applied in an averaged manner, in which case an average treatment pressure is used to determine a constant fracture height (Rahim and Holditch, 1993). In *complex modeling* of height growth, the concept is applied for every time instant at every lateral fracture location, as depicted in Figure 10-2. This is also the basis for pseudo-3D fracture modeling.

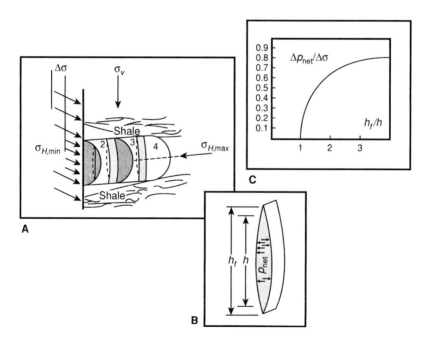

FIGURE 10-2. Fracture geometry and height growth.

LOGGING METHODS AND TRACERS

Once a reservoir interval has been fracture treated, any number of logging methods are available to evaluate the created fracture. The most widely used methods include gamma ray, spectral gamma ray, temperature, production, full wave-form sonic, and oriented gamma ray logging. Spectral gamma ray images use multi-isotope tracers to identify fracturing outcomes such as (1) propped versus hydraulic fracture height at the wellbore, (2) proppant distribution at the wellbore, (3) perforations or target intervals that were not stimulated, and (4) fracture conductivity as a function of fracture width and proppant concentration.

Mutually distinctive tracers can be applied in sequential fracture stages to determine the *staging efficiency* of an acid or propped fracture treatment. If the stress or pore pressure contrast between reservoir layers is larger than anticipated, a single-stage treatment may result in inefficient coverage. Conversely, radioactive tracers may indicate that multiple stages are not necessary, in which case subsequent treatments can employ fewer stages. The effectiveness of gel, foam, or mechanical diverters can be determined using distinctive tracers in the various treatment stages. Tracers also can establish whether or not ball sealers are effective in distributing treatment fluids over an entire interval.

The use of radioactive tracers is recommended when one or more of the following applies:

■ Thick intervals of reservoir are to be stimulated (e.g., greater than 45 ft).

■ Stress contrast between the zone of interest and adjacent barriers is low, (e.g., less than 700 psi).

■ A limited entry perforating strategy is planned.

■ Specialty proppants will be used, especially if "tailed-in" at the end of the treatment.

■ Fluid leak-off is unknown or expected to be higher than usual.

Temperature logging can determine post-treatment hydraulic fracture height and fluid distribution at the wellbore, but is not indicative of proppant placement or distribution. Cold fluids (ambient surface temperature) injected into the formation can be detected readily by a change in the temperature profile within a wellbore. A series of logging passes is usually sufficient to determine the total treated height. Intervals that received a large volume of injected fluids and/or proppant will require a much longer time to return to thermal equilibrium.

A WORD ON FRACTURE MAPPING

A powerful new category of direct fracture diagnostic techniques has taken shape over the last decade or so that includes various microseismic and tiltmeter fracture mapping techniques (cf. the work of Vinegar, et al., 1992). Fracture mapping relies on measurement of acoustic signals and rock deformation caused by the fracture process to determine created fracture geometry.

The hydraulic fracture process can be viewed as a series of mini-earthquakes. An extensive set of distinct acoustic signals is generated as the rock is stressed and deformed. In principle, by monitoring and mapping these microseismic events and deformations, the evolution and extent of the fracture can be established directly. These techniques hold great promise over conventional direct measurement techniques, such as radioactive tracers or temperature logs, as the depth of investigation is virtually unlimited—effectively allowing us to monitor events tens or even hundreds of feet from the treatment well.

When fixed downhole geophone arrays (borehole seismics) are used to monitor microseismic events created by the fracturing process, this is known as *passive seismic imaging*. *Active seismic imaging* or *cross-well tomography* implies the systematic transmitting and receiving of a series of acoustic signals across a fracture plane in order to establish fracture extent (Figure 10–3). Though cross-well surveys have been performed by many companies, further advances are still needed in the areas of source creation, data acquisition, and interpretation methods.

Fracture mapping with tiltmeters has been used extensively over the past decade (Fisher, 2001), notwithstanding a range of applications that is somewhat limited. At a basic level, deformations caused by the hydraulic fracture treatment are transmitted far from the well by surrounding rock strata. In the case of relatively shallow formations (up to several thousand feet), this deformation results in a "tilt" that is easily measured on the surface. Modern tiltmeters—essentially equivalent to the level a carpenter uses, but much more sensitive—

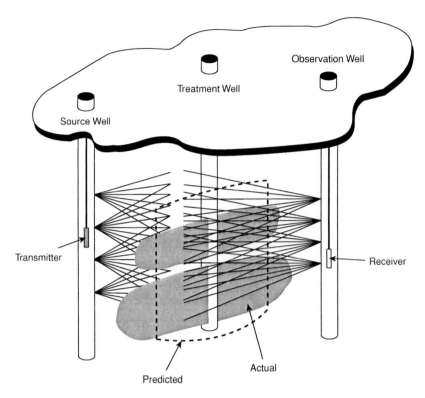

FIGURE 10-3. Seismic imaging is a powerful fracture diagnostic technique.

are capable of measuring deviations of just 0.0000001 percent. *Surface tiltmeters* are particularly useful for determining fracture orientation in shallow formations. *Downhole tiltmeters* are primarily used to determine fracture height and length. Their use has been severely limited by the need for offset observations wells in which to position the tiltmeters.

The latest generation of downhole tiltmeters can be placed directly in the treatment well. With time, this should broaden the use and utility of the technique.

WELL TESTING

In low permeability formations, a well test before the fracturing treatment is not practical, so knowledge of the permeability is typically limited. In this case, a buildup test in the newly fractured well is intended to obtain the permeability and the fracture extent, simultaneously. Unfortunately, this is an ill-posed problem in the sense that many different combinations of the two unknowns may provide a plausible solution. In high permeability formations, the permeability is usually known ahead of time, and the primary goal of a post-treatment test is to evaluate the created fracture.

For well-testing purposes, an infinite acting reservoir can be considered. The transient behavior of a vertical well intersected by a finite conductivity fracture was aptly described by Cinco-Ley and his co-workers (1978, 1981). Figure 10-4 shows the log-log *diagnostic plot* of dimensionless pressure versus dimensionless time and parameterized by the dimensionless fracture conductivity.

In the *bilinear* flow regime, where the flow is determined by both the reservoir and fracture properties, the dimensionless pressure can be expressed as

$$p_D \approx \frac{\pi}{\Gamma(5/4)\sqrt{2C_{fD}}} t_{Dxf}^{1/4} \tag{10-1}$$

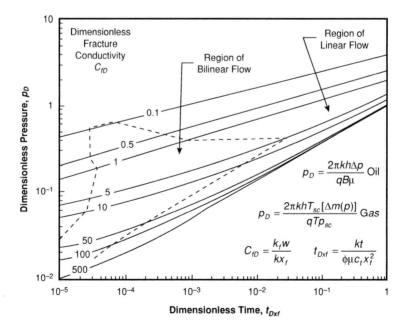

FIGURE 10-4. Log-log diagnostic plot for a fractured vertical well.

where t_{Dxf} is the dimensionless time and fracture half-length is the characteristic dimension. Accordingly, this flow regime is distinguished by a quarter-slope on the log-log pressure and pressure derivative plot.

Once such a regime is identified on a well-test diagnostic plot, a *specialized plot* of the pressure versus the quarter-root of time can be constructed. The slope, m_{bf}, of the (fitted) straight line is a combination of the reservoir and fracture properties:

$$m_{bf} = \frac{\pi}{2\sqrt{2}\Gamma(5/4)\sqrt{C_{fD}}} \frac{B\mu q}{\pi kh} \left(\frac{k}{\phi\mu c_t x_f^2}\right)^{1/4} \tag{10-2}$$

$$= \left(0.390 \frac{B\mu^{3/4}q}{h\phi^{1/4}c_t^{1/4}}\right) \frac{1}{k^{1/4}k_f^{1/2}w^{1/2}}$$

It can be used to obtain one or the other quantity, or their combination, depending on the available information. As is obvious from the above equation, the formation permeability and the fracture conductivity cannot be determined simultaneously from this regime. Knowing the formation permeability, the fracture conductivity ($k_f \times w$) can be determined from the slope, but the fracture extent cannot.

Our suggestion for a properly designed and executed treatment is to assume a dimensionless fracture conductivity, $C_{fD} = 1.6$, then determine an equivalent fracture conductivity from Equation 10-2, and calculate an equivalent fracture length:

$$x_{feq} = \left(\frac{0.308}{m_{bf}} \frac{Bq}{h} \frac{\mu^{3/4}}{k^{3/4}c_t^{1/4}\phi^{1/4}}\right)^2 \tag{10-3}$$

Comparing the equivalent fracture length to the design length may provide valuable information on the success of the fracturing job.

The actual fracture extent might also be determined from the subsequent *formation linear* or late-time *pseudo-radial* flow regimes. Unfortunately, the formation linear flow regime is often too limited in duration to be distinguishable, and the pseudo-radial flow regime may not be available owing to boundary effects.

In formation linear flow, the approximate solution is:

$$p_i - p_{wfs} = \frac{2\pi q\mu B}{kh} \sqrt{\pi} \left(\frac{kt}{\phi\mu c_t x_f^2}\right)^{1/2} \tag{10-4}$$

Therefore, the fracture half-length can be obtained from the slope of a specialized plot of pressure versus square-root of time, according to:

$$x_f = \frac{11.14}{m_{flf}} \frac{qB}{h} \left(\frac{\mu}{\phi c_t k} \right)^{1/2}$$ (10-5)

This flow regime (if it exists) is not influenced by the fracture conductivity.

In the literature, several other effects are considered: influence of boundaries, reservoir shape and well location, commingled reservoirs, partial vertical penetration, non-Darcy flow in the fracture and/or the formation, permeability anisotropy, double porosity, phase changes, fracture-face damage, and spatial variations in fracture conductivity.

EVALUATION OF HPF TREATMENTS—A UNIFIED APPROACH

Production Results

The evaluation of high permeability fracture treatments can be viewed on several different levels, the most pragmatic (and most common) being economic justification (i.e., production results). Simply put, HPF has gained widespread acceptance because it allows operators to make more oil at less cost. McLarty and DeBonis (1995) reported that *fracpack* treatments typically result in production increases of 2 to $2\frac{1}{2}$ times that of comparable gravel packs, and offered the example cases shown in Table 10-1.

Similar reports of production increase are scattered throughout the body of HPF literature. Stewart et al. (1995) presents a relatively

TABLE 10-1. Example HPF Production Results

Job Type	Before	After
New Well Comparison	460 bopd	1,216 bopd
Recompletion (oil)	1,300 bopd	2,200 bopd
Recompletion (gas)	3.8 MMcfd	13.2 MMcfd
Sand Failure	200 bopd	800 bopd

Source: McLarty and DeBonis (1995).

comprehensive economic justification for HPF that considers (in addition to productivity improvements) the incremental cost of HPF treatments and the associated payouts, operating expenses, relative decline rates, and reserve recovery acceleration issues.

Evaluation of Real-Time HPF Treatment Data

There is increasing recognition of the value of real-time HPF treatment data. Complete treatment records and digital treatment datasets are now routinely collected and evaluated as part of post-treatment analysis.

Treatment reconstruction and post mortem diagnosis hold tremendous potential to improve HPF design and execution, but the usefulness of many ongoing efforts in this regard is limited. The proliferation of user friendly, black box software and simulations has often obscured instead of improved the physical understanding of the process.

The practice of evaluating real-time datasets by *net pressure history matching* is often suspect. Complexities incorporated in a 3D simulator to improve the "match" unwittingly sacrifice the uniqueness (usefulness) of the evaluation, and thereby destroy the predictive capability of the simulation. These activities provide operators with little more than qualitative direction on a case-by-case basis.

Contrasting this approach, consider the step-wise approach for evaluation of bottomhole treating pressures as outlined by Valkó et al. (1996):

1. A leakoff coefficient is determined from an evaluation of minifrac data using a minimum number of assumptions, minimum input data, and minimum user interaction. Radial fracture geometry and a combined Nolte-Shlyapobersky method are suggested.
2. Using the obtained leakoff coefficient, an almost automatic procedure is suggested to estimate the created fracture dimensions and the areal proppant concentration from the bottomhole pressure curve monitored during the execution of the HPF treatment. This procedure (termed "slopes analysis") is further developed in a separate section below as a fundamental and important building block for evaluation of real-time HPF data.
3. The obtained fracture dimensions and areal proppant concentration are converted into an equivalent fracture extent and conductivity. The actual performance of the well is analyzed using well test procedures, and these results are compared to the results of the slopes analysis.

4. Conducting the procedure above for a cross-section of treatments in a given control volume results in a data bank that improves the predictability and outcome of HPF treatments.

At present, there seems to be a trend in the industry to support joint efforts and assist mutual exchange of information. The procedure above provides a coherent (though not exclusive) framework to compare HPF data from various sources using a common, cost-effective evaluation methodology.

Post-Treatment Well Tests in HPF

For post-treatment evaluation, temperature logs and various fracture mapping techniques have gained increasing importance. However, from the standpoint of future production, by far the most important is pressure transient analysis. While avoiding an exhaustive treatment of the subject, it is appropriate at this juncture to address several issues related to pressure transient analysis in HPF wells, especially positive skin factors, which pose the largest challenge to treatment evaluation.

The performance of a vertically fractured well under pseudo-steady-state flow conditions was investigated by McGuire and Sikora (1960) using a physical analog (electric current). A similar study for gas wells was conducted by van Poollen et al. (1958). For the "unsteady-state" case, a whole series of works was initiated by Gringarten and Ramey (1974), and continued by Cinco-Ley et al. (1978). They clarified concepts of the infinite-conductivity fracture, uniform-flux fracture, and finite-conductivity fracture. From the formation perspective, double-porosity reservoirs, multilayered reservoirs, and several different boundary geometries have been considered. The typical flow regimes (fracture linear, bilinear, pseudo-radial) have been well documented in the literature and were discussed above. Deviations from ideality (non-Darcy effects) have also been considered.

Post-treatment pressure transient analysis for HPF wells starts with a log-log diagnostic plot, including the pressure derivative. Once the different flow regimes are identified, specialized plots can be used to obtain the characteristics of the created fracture. In principle, fracture length and/or conductivity can be determined using the prior knowledge of permeability. For HPF, however, the relatively large arsenal of fractured well, pressure transient diagnostics, and analysis has proven somewhat ineffective. Often it is difficult to reveal the marked characteristics of an existing fracture on the diagnostic plot. In fact, the well often behaves similar to a slightly damaged, unstimulated well. An HPF treatment is often considered successful if a large positive skin of order +10 or more is decreased to the range of +1 to +4. These (still) positive skin factors create the largest challenge of treatment evaluation.

The obvious discrepancy between theory and practice has been attributed to several factors, some of which are well documented and understood and some others of which are still in the form of hypotheses:

- *Factors causing a decrease of apparent permeability in the fracture.* The most familiar factor that decreases the apparent permeability of the proppant pack, and therefore fracture conductivity, is proppant pack damage. The reduction of permeability caused by residue from the gelled fluid and proppant crushing are well understood. Since these phenomena exist in any fracture, they cannot be the general cause of the discrepancy in high permeability fracturing. Non-Darcy flow in the fracture is also reasonably well understood. Separation of rate-independent skin from the variable-rate component by multiple-rate well testing is a standard practice. The effect of phase change in the fracture is less straightforward to quantify.

- *Factors decreasing the apparent width.* Embedment of the proppant in a soft formation is now well documented in the literature (e.g., Lacy et al., 1996).

- *Fracture-face skin effect.* The two sources of this phenomenon are filter cake residue and polymer invaded zones. Sometimes the long term clean-up (decrease of the skin effect) of a stimulated well is considered as indirect proof of such damage. It is taken that linear polymer fluids invade more deeply into the formation and hence cause more fracture-face damage, as discussed by Mathur et al. (1995).

- *Permeability anisotropy.* While the anisotropy of permeability has only a limited effect on pseudo-radial flow, the early time transient flow regime of a stimulated well is very sensitive to anisotropy. This fact is often neglected when characterizing the well with one single skin effect.

- *Concept of skin.* It must be emphasized that the concept of negative skin as the only measure of the quality of a well might be a source of the discrepancy itself.

Validity of the Skin Concept in HPF

There is, in fact, no clear theoretical base for obtaining a negative skin from short-time well test data—which is distorted by wellbore storage if the well has been stimulated. In this case, a classic type-curve analysis that assumes an infinite acting reservoir, wellbore storage, and skin effect is not based on sound physical principles and might cause unrealistic conclusions.

The validity of the pseudo-skin concept during the transient production period is also an important issue. In general, the pseudo-skin concept is valid only at late times. Thus, a fracture designed for late-time performance may not perform optimally at early times. The penalty in initial production rate associated with optimizing fracture dimensions for late time has not been investigated. Yet, it is reasonable to assume that the loss in performance is minimal for high permeability reservoirs where the dimensionless times corresponding to a month or a year are much larger than for low permeability reservoirs.

SLOPES ANALYSIS

Complete tip screenout is expected to produce a distinct behavior in the treating pressure; that is, the treating pressure should markedly increase with time. However, HPF treatments often exhibit numerous increasing pressure intervals that are interrupted by anomalous pressure decreases, most probably because fracture extension can still occur from time-to-time (i.e., in many cases, a single complete tip screenout is not achieved).

Work presented by Valkó, Oligney, and Schraufnagel (1996) provides a simple tool for examining such behavior. Treating pressure curves are analyzed to gain insight to the evolution of fracture extent and a plausible end-of-job proppant distribution.

In developing the tool, several design parameters were intentionally imposed: the method should require minimum user input beyond the real treatment data, it should be relatively independent of the fracture propagation model used, and it should not be a history-matching procedure. In accordance with the basic requirement of model independence, the slopes analysis method is a screening tool based on simple equations and a well-defined (re-constructible) algorithm. Based on its simplicity, the tool lends itself to real-time use as well.

Assumptions

During tip screenout, the fracture width is inflated while the area of the fracture faces remains theoretically constant. This phenomenon should manifest itself by a marked increase in the treating pressure. In practice, the increasing pressure intervals may be interrupted by an anomalous pressure decrease because fracture extension can still occur from time-to-time. Based on this rationale, the HPF treatment is considered a series of (regular) arrested extension/width growth intervals interrupted by (irregular) fracture area extension intervals.

In this case, the treatment can be decomposed into sequential periods of constant fracture area separated by periods (possibly several) of fracture extension. The time periods are located by a simple processing of the treatment pressure curve.

If this vision of the treatment is accepted, then the slope of the increasing pressure curve during a width inflation period may be interpreted to obtain the "packing radius" of the fracture at that point during the treatment (i.e., characteristic for the given period). Putting together a sequence of packing-radii estimates gives a scenario that—combined with additional information on the proppant injection history—in turn yields the final proppant distribution.

In transforming the idea to a working algorithm, several assumptions must be made, both on fracture geometry and the character of the leakoff process. The following assumptions are employed:

1. The created fracture is vertical with a radial geometry;
2. Fluid leakoff can be described by the Carter leakoff model (Howard and Fast, 1957) in conjunction with the power-law type area growth used by Nolte (1979), or by the one of the detailed leakoff models discussed in Chapter 5;
3. Fracture packing radius may vary with time, being allowed to increase or decrease;
4. Hydraulic fracture radius (which defines leakoff area) cannot decrease and is the maximum of the packing radii that have occurred up to the given time;
5. During regular width-inflation periods, the pressure slope is defined by linear elastic rock behavior and fluid material balance with friction effects being negligible; and
6. Injected proppant is distributed evenly along the actual packing area during each incremental period of arrested extension/width growth.

The suggested method consists of several steps. First, those portions of the bottomhole pressure curve are selected that show positive slope. The slope is then interpreted assuming that the pressure increase is caused by width inflation. The interpretation results in a packing radius that corresponds to a given time point. A step-by-step processing of the entire curve gives a history of the packing radius, though still not providing information on those intervals when the slope is negative. The history is made complete by interpolating between the known values.

Based on this history of packing radius evolution, the final proppant distribution is easily determined by superimposing real-time proppant injection data. Final proppant distribution (which implies fracture length and width) is the practical result of the proposed slopes analysis.

Restricted Growth Theory

Tip screenout can be considered as inflating the fracture width while the area of the fracture face does not increase. If the average width is denoted by w and the fracture-face area (one wing, one face) by A, then

$$\frac{dw}{dt} = \frac{1}{A}(i - q_L) \tag{10-6}$$

where i is the injection rate (per one wing) and q_L is the fluid-loss rate (from one wing).

The basic notation is shown in Figure 10-5. Assuming that the fracture is radial with radius R, then

$$A = \frac{\pi R^2}{2} \tag{10-7}$$

As a first approximation, assume that the pressure in the inflating fracture does not depend on location (i.e., it is homogeneous). The net pressure, for example the excess pressure above the minimum principal stress, is directly proportional to the average width:

$$p_n = \frac{3\pi E'}{16R} w \tag{10-8}$$

where E' is the plane-strain modulus (see Chapter 4).

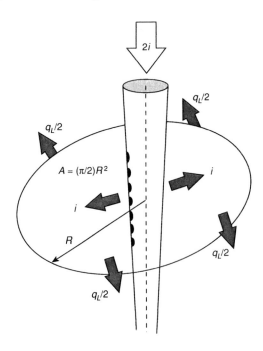

FIGURE 10-5. HPF radial fracture geometry.

Substituting Equations 10-7 and 10-8 into Equation 10-6, the time derivative of net pressure is obtained as

$$\frac{dp}{dt} = \left(\frac{3\pi E'}{16R}\right)\left(\frac{2}{\pi R^2}\right)(i - q_L) \tag{10-9}$$

where the subscript for net pressure is dropped because the derivative of bottomhole pressure and that of net pressure are equal.

Recording bottomhole pressure and injection rate provides the possibility of using Equation 10-9 to determine R. For this purpose, an estimate of q_L is needed.

Details of the Carter leakoff model are given in Chapter 4. Assuming that the fracture has extended up to the given time t according to Nolte's power-law assumption and is arrested at the given time instant t, the leakoff rate $q_{L,t}$ immediately after the arrest is given by

$$q_{L,t} = 2AC_L \frac{1}{\sqrt{t}}\left(\frac{\partial g(\Delta t_D, \alpha)}{\partial \Delta t_D}\right)_{\Delta t_D = 0} \tag{10-10}$$

where A is the current fracture area and α is the power law exponent of the areal growth. The two-variable g-function was discussed in Chapter 4.

For a radial fracture created by injecting a Newtonian fluid, the exponent is taken as $\alpha = 8/9$ and the derivative of the g-function is

$$\left[\frac{\partial g(\Delta t_D, 8/9)}{\partial d\Delta t_D} \right]_{\Delta t_D = 0} = 1.91 \tag{10-11}$$

Therefore, the estimate of leakoff rate is obtained as

$$q_{L,t} = 2AC_L \frac{1}{\sqrt{t}} 1.91 \tag{10-12}$$

Slopes Analysis Algorithms

The restricted growth theory is combined with simple material balance computations to form the slopes analysis method as demonstrated below using a sample set of HPF data provided by Shell E&P Technology Company.

Selecting Intervals of Width Inflation

Figure 10-6 is the bottomhole pressure recorded during a HPF treatment. While it may look "not typical," the fact is that most of the data sets available (without the natural self-censoring of publishing authors) are atypical in one or more respects. The recommended approach is based exactly on this premise (i.e., avoiding premature assumptions about the form of the pressure curve). The slopes analysis approach can be better described as a signal processing operation than one of fitting a given model to the data.

The suggested method consists of selecting those portions of the bottomhole pressure curve that show positive slope. Straight lines are fitted to the points corresponding to each such interval. Using a simple algorithm, one can select points satisfying the criterion of restricted fracture growth. Straight lines are fitted to the individual series to arrive at the plot shown in Figure 10-7.

The slope of the straight line gives an average pressure derivative corresponding to the given time interval of restricted growth. In view of the stated assumptions, these slopes contain information that defines

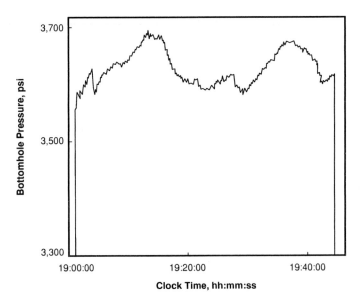

FIGURE 10-6. Bottomhole treating pressure from HPF treatment.

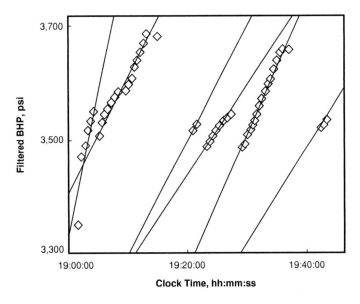

FIGURE 10-7. Bottomhole pressure points corresponding to width inflation intervals and corresponding "straight lines."

the actual packing radius corresponding to discrete moments during the HPF treatment.

Determining the Packing Radius Corresponding to a Width Inflation Period

Substituting the obtained expression for the leakoff rate, Equation 10-9 can be rewritten as

$$m = \left(\frac{3\pi E'}{16R}\right)\left(\frac{2}{\pi R^2}\right)\left[i - 2\left(\frac{\pi R^2}{2}\right)C_L \frac{1}{\sqrt{t}}1.91\right] \tag{10-13}$$

Rearranging, we obtain

$$R^3 + R^2\left(\frac{2.25E'C_L}{m\sqrt{t}}\right) - \left(\frac{0.375E'i}{m}\right) = 0 \tag{10-14}$$

Once a restricted-growth interval is selected, knowing the slope m, and the injection rate i, at a given time t, Equation 10-14 can be solved for R. Since the equation is cubic, an explicit solution can be given.

Equation 10-14 can be used with the actual one-wing slurry injection rate, i, recorded at time t and the slope of the pressure versus time curve at that instant. The obtained solution is the *packing radius*. Figure 10-8 shows the packing radius obtained from recorded data of the example HPF treatment. As seen from the figure, after a certain period of pumping time (approximately 25 minutes), the packing radius begins to decrease. In other words, near the end of the treatment only the near-wellbore part of the fracture was "packed." This is consistent with the treatment objectives, and was achieved by gradually decreasing the injection rate at the final stages of the treatment.

Interpolation Between Known Values of the Packing Radius

Since the packing radius is obtained only in those selected intervals where width-inflation can be assumed, a simple tool is needed to fill in the gaps. A simple logarithmic interpolation is used to estimate the packing radius between the known values.

In addition, one can estimate the "hydraulic" fracture radius at time t as the maximum of the packing radii up to that point (see dashed line in Figure 10-8). While proppant is placed within the actual

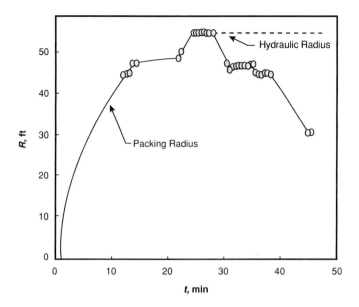

FIGURE 10-8. Estimated packing radius with interpolation.

packing radius, leakoff takes place along the area determined by the hydraulic fracture extent. Knowledge of the hydraulic fracture extent is useful for further material balance considerations.

Determining the Final Areal Proppant Concentration

Final proppant concentration (proppant distribution) in the fracture can be derived in a relatively straightforward fashion from the packing radius curve and knowledge of the bottomhole proppant concentration as a function of time. The standard job record typically includes this information.

Calculation of the final areal proppant concentration in the fracture follows a simple scheme:

1. For every time interval, Δt, determine the mass of proppant entering the fracture.
2. Assume this mass to be uniformly distributed inside the packing radius corresponding to the given time step.
3. Obtain the mass of proppant in a "ring" between radius R_1 and R_2 by summing up (accumulating) the mass of proppant placed during the whole treatment.

4. Repeat Step 3 for all rings to obtain the areal proppant concentration as a function of radial location R.

Application of the scheme above to the example data results in the areal proppant concentration as a function of the radial distance from the center of the perforations, R. The areal proppant concentration distribution for the example dataset is shown in Figure 10-9.

The demonstrated method for evaluating pressure behavior of HPF treatments is not based on specific fracture mechanics and/or proppant transport models. Rather, it takes the pressure curve "as is" and processes it using minimum additional data. The usual data records of a job (slurry injection rate, bottomhole proppant concentration, and bottomhole pressure) can be used to estimate fracture extent and the distribution of proppant in the fracture. The only other additional input parameters necessary for the analysis are plane-strain modulus and leakoff coefficient.

The success of this procedure depends on the validity of the key assumption that positive slopes observed in the bottomhole pressure curve are caused by restricted fracture extension/width growth. If there

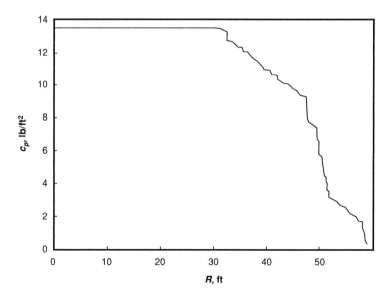

FIGURE 10-9. Final areal proppant concentration as a function of radial distance from the center of the perforations.

is no time interval satisfying the criterion of restricted extension or other phenomena involved mask the effect (e.g., pressure transients caused by sharp changes of the injection rate or dramatic changes in friction pressure resulting from proppant concentration changes), the estimated packing radius might be in considerable error. Nevertheless, the suggested procedure is considered a substantive step in the analysis of HPF treatment pressure data.

Nomenclature

A	area, m, ft
A_0	fracture area at TSO
B	formation volume factor
b_1	radial damage radius, m, ft
b_2	fracture-face damage depth, m, ft
b_N	intercept, Nolte method, Pa, psi
B_o	oil formation volume factor, RB/STB
b_s	depth of polymer invasion, ft
c	line crack half-length, m, ft
C_A	well-reservoir shape factor (single well)
C_{fD}	dimensionless fracture conductivity
C_L	Carter leakoff coefficient (apparent, with respect to fracture area), $m/s^{1/2}$, $ft/min^{1/2}$
$C_{L,p}$	Carter leakoff coefficient (with respect to permeable layer), $m/s^{1/2}$, $ft/min^{1/2}$
C_w	wall component of leakoff coefficient, $m/s^{1/2}$, $ft/min^{1/2}$
c_t	total reservoir compressibility, Pa^{-1}, psi^{-1}
E	Young modulus, Pa, psi

E'	plane strain modulus, Pa, psi
h_f	fracture height, m, ft
h_p	permeable height, m, ft
i	injection rate per one wing, $m^3 \cdot s^{-1}$, BPM
I_x	penetration ratio: $2x_f/x_e$
J	productivity index, $m^3 \cdot s^{-1} \cdot Pa^{-1}$, STB/d/psi
J_D	dimensionless Productivity Index
k	reservoir permeability, m^2, md
k_1	permeability in the equivalent radial damaged zone, m^2, md
k_2	permeability in the zone of fracture-face invasion (outside the radial damage zone), m^2, md
k_3	permeability in the zone of fracture-face invasion (inside the radial damage zone), m^2, md
k_f	fracture permeability, m^2, md
k_r	reservoir permeability, m^2, md
K	fluid consistency index, $Pa \cdot s^n$, lb_f s^n/ft^2
k_f	fracture permeability, md
k_{fg}	relative permeability of ga
K_L	opening-time distribution factor, dimensionless
k_s	damaged permeability, md
M_{tso}	total sand weight
m_N	slope, Nolte method, Pa, psi
m_{bf}	slope, bilinear flow specialized plot, $Pa \cdot s^{-1/4}$, psi $hr^{1/4}$
n	flow behavior index, dimensionless
N_p	proppant number
p_i	initial reservoir pressure, Pa, psi
p_e	outer boundary constant pressure, Pa, psi
$\bar{p}$	average reservoir pressure, Pa, psi
p_{wf}	flowing bottomhole pressure, Pa, psi
p_c	closure pressure, Pa, psi
p_D	dimensionless pressure
$\Delta p(t_0)$	net pressure at TSO
q	production rate (standard conditions), $m^3 \cdot s^{-1}$, STB/d

q_g gas flow rate, STB/D

q_L leakoff rate (1 wing, through 2 faces) m^3/s, BPM

r distance from fracture tip, m, ft

r_e equivalent reservoir radius, m, ft

R_f radius of a radial fracture, m, ft

R_o filter-cake resistance, $Pa \cdot s \cdot m^{-1}$

r_p ratio of permeable to fracture area

r_w wellbore radius, m, ft

s total (pseudo) skin of well

s_d effective skin factor due to radial damage and fracture-face damage

s_f skin accounting for finite conductivity fracture with no fracture-face skin and no radial damage

s_{fs} fracture face skin

S_g gas saturation

s_{ND} skin due to non-Darcy flow

S_h minimum horizontal stress, Pa, psi

S_v absolute vertical stress, Pa, psi

S_f fracture stiffness, $Pa \cdot m^{-1}$, psi ft^{-1}

S_p spurt loss coefficient (apparent), m, ft

$S_{p,p}$ spurt loss coefficient (with respect to permeable layer), m, ft

s_t total composite skin

S_w water saturation

t time, s, hr

T temperature, K, °R

t_0 total time to TSO

t_D dimensionless time (with respect to well radius)

t_{Dxf} dimensionless time (with respect to fracture half-length)

t_e time at end of pumping, s, min

V volume of one fracture wing, m^3, ft^3

$V_F(t_0)$ fracture volume at TSO

V_i volume of injected fluid into 1 wing, m^3, ft^3

V_p volume of proppant in pay, ft^3

V_r volume of pay, ft^3

w	average fracture width, m, ft
w_e	average fracture width at end of pumping, m, ft
w_p	average propped width, m, ft
x_f	fracture half-length, m, ft
x_e	size of study area in x-direction
y_e	size of study area in y-direction
Z	gas compressibility factor
α	exponent of fracture area growth, dimensionless
α	Biot's poroelastic constant, diminsionless
α	parameter of the Fan-Economides model
α_1	conversion factor (for field units 887.22)
ε	proppant schedule exponent (also pad fraction), dimensionless
ϕ	porosity, dimensionless
ϕ_p	proppant pack porosity, dimensionless
γ	ratio of average to maximum width, dimensionless
$\dot{\gamma}$	shear rate, 1/s
η	fluid efficiency, dimensionless
η	parameter of the Fan-Economides model
η_e	fluid efficiency at end of pumping, dimensionless
μ	viscosity, Pa·s, cp
μ_a	apparent viscosity, Pa·s, cp
μ_e	equivalent Newtonian viscosity, Pa·s, cp
μ_r	viscosity of reservoir fluid, Pa·s, cp
ν	Poisson ratio, dimensionless
σ	interfacial tension (pressure units)
σ_h	effective horizontal stress, Pa, psi
σ_v	effective vertical stress, Pa, psi
τ	shear stress, Pa, psi
ϕ	porosity
Γ	Euler's Constant = 0.57721566 ...

Glossary

acidizing: the stimulation of oil or gas production by injecting a solution of hydrochloric or other acid into a producing formation.

anionic: negatively charged, characterized by a surface active anion.

batch mix: fluid for use in a fracturing job that has been fully prepared in the fluid storage tank before the start of the pumping operation.

bauxite: an aluminum oxide used as a proppant in deep high pressure zones.

bottomhole pressure: pressure on the bottom of the well, which can be hydrostatic or a combination of hydrostatic and applied pressures.

breaker: an enzyme, oxidizing agent, or acid added to a fracturing fluid to degrade or "break" the polymer, dramatically reducing fluid viscosity and aiding in fracture closure and cleanup.

buffer: a weak acid (acetic, formic, or fumaric) used to reduce the pH of a fluid, or a base (e.g., sodium bicarbonate or sodium carbonate) used to maintain a high pH range.

bullheading: loading a well with acid or fluid, without propping materials, for the purpose of breaking down the formation.

cation: positively charged ion.

centipoise (cp): the unit of measurement for viscosity, equal to 1/100th of a poise. (Water has a viscosity of 1 cp; olive oil has a viscosity of 100 cp or 1 poise.)

coning: the encroachment of reservoir water into the oil column and/ or producing well because of excessive pressure drawdown.

connate water: non-producible water retained in the pore spaces of a formation—viewed as having occupied the rock interstices from the time the formation was created, often expressed as a percentage of the total pore space available.

crosslinker: a chemical added to fracturing fluid that effectively "links" parallel chains in the polymer, resulting in a complex molecule and increased fluid viscosity.

Darcy: a unit of measurement for permeability.

density: the weight of a unit volume of a substance in pounds per gallon or pounds per cubic centimeter; for example, since 1 cubic centimeter of water weighs 1 gram, its density is 1 gram per cubic centimeter. (See also *specific gravity.*)

differential pressure: the pressure difference between two sources that meet at an interface.

dilatent fluid: a fluid that exhibits no yield stress but for which the slope of the rheological curve increases with increasing shear rate. (Compare to *pseudoplastic fluid* and *Newtonian fluid.*)

drainage radius: one-half the distance between properly spaced wells, or otherwise the no-flow boundary at the edge of a reservoir.

drawdown: the difference between static and flowing bottomhole pressures.

ellipsoid flow: the laminar fluid flow regime that occurs in an elliptical cross section when the ratio of fracture length to fracture height is extremely large, corresponding to the PKN geometry.

emulsion: the suspension of a very finely divided oily or resinous liquid in another liquid, or vice versa, as compared to a solution that is a uniform mixture of two or more substances. (Of particular concern in hydraulic fracturing is emulsions created between treating fluids and oil in the formation, which can block the natural formation permeability.)

encapsulated breaker: a breaker wrapped or "encapsulated" in a soluble coating that dissolves slowly downhole (i.e., to intentionally delay the degrading action of the breaker).

enzyme breaker: an efficient chemical breaker that can be employed when bottomhole fracture treating temperatures are between 60°F to 200°F (pH less than 10).

flow capacity: the product of formation permeability (in millidarcies) and formation thickness (in feet).

fluid density: the weight of a fluid expressed in pounds per square inch or pounds per gallon.

fluid flow: the motion of a fluid, described more particularly by the type of fluid (e.g., Newtonian, plastic, pseudoplastic, dilatant), fluid properties (e.g., viscosity and density), the geometry of the system or flow channel, and the flow velocity.

fluid friction: the friction, expressed as a pressure loss on top of the useful work to be done, which results from fluid flow through surface equipment and downhole tubulars. (Fluid friction must be considered when determining pressure needs and power requirements to fracture treat a well.)

fluid loss: the amount of fluid that escapes or "leaks off" from the created fracture into the formation during a fracture treatment. (Knowledge of fluid loss is necessary to determine fracture dimensions; to some extent, fluid loss can be controlled to meet treatment objectives.)

fluid loss additive: a chemical used to reduce fluid loss during a fracture treatment, thereby increasing fluid efficiency (maximizing the fracture dimensions created with a given fluid volume) and reducing the potential for damage to the formation.

force: pressure times area.

formation volume factor: the reservoir volume occupied by a unit volume of oil at standard surface conditions, including dissolved gas.

fracture: a parting or crack in a formation (noun); or to part or create a crack in a formation (verb).

fracturing: the use of a special fluid under hydraulic pressure to cause a parting or "fracture" in a formation.

friction pressure: the pressure or force (generally undesirable) caused by the motion of a fluid against a surface, such as oilfield tubulars or surface equipment.

gradient: the unitized rate of increase or decrease in a parameter of interest, such as temperature or pressure.

horizontal fracture: a fracture that is oriented parallel to the surface of the earth, generally not occurring at depths greater than 1500 ft.

hydraulic fracturing: a method of stimulating production (or injection) in which fractures are induced by applying very high fluid pressure to the face of the formation.

hydraulic horsepower: a rate of work measure commonly used to define the output capability of positive displacement plunger-type pumps employed in hydraulic fracturing.

hydrocarbons: organic compounds comprised of hydrogen and carbon, commonly existing in three forms or phases: coal (solid), oil (liquid), and natural gas (vapor).

hydrostatic head: see *hydrostatic pressure.*

hydrostatic pressure: the pressure exerted at the base of a column of liquid, such as standing in a well, which is dependent on the fluid density, or "weight." (Fresh water exerts a hydrostatic head of 0.433 psi per foot of height. The hydrostatic head of any liquid may be determined, or modified, by considering its specific gravity relative to that of water.)

inflow performance: the explicit relationship that exists between the flow rate from a formation and the flowing pressure at the bottom of the well, typically presented in Cartesian coordinates.

in-situ stress: the state of stress within a formation, which determines fracture orientation and treating pressures. (See also *stresses at depth.*)

instantaneous shut-in pressure (ISIP): the wellhead pressure at the very moment the frac pumps are shut down (while less common, may also be used in reference to bottomhole pressure readings).

interfacial tension: the force acting at an interface between two liquids or between a liquid and a solid. (See also *surface tension.*)

ionic: having an electric charge, either positive or negative.

laminar flow: the motion of a fluid in layers or laminae that are at all times parallel to the direction of flow. (See also *turbulent flow*.)

leakoff: see *fluid loss*.

limited entry perforating strategy: the use of a very limited number of carefully sized perforations to create a self-limiting critical flow condition (flow rate reaches a maximum, independent of pressure differential) in order to distribute a fracture treatment over multiple zones in a thick formation.

matrix acidizing: a method of treating the near wellbore formation with acid to improve permeability without fracturing.

maximum horizontal stress: the larger of the two horizontal principal stresses, orthogonal to and larger than the minimum horizontal stress because it includes additional horizontal stress components related to tectonic phenomena. (See also *stresses at depth*.)

minimum horizontal stress: the smaller of the two horizontal principal stresses, resulting from vertical-to-lateral translation of the overburden stress through the Poisson ratio relationship. (See also *stresses at depth*.)

Newtonian fluid: a fluid that exhibits no yield stress (flow is initiated immediately under an infinitesimal shear stress) and straight-line rheological behavior (shear stress varies linearly with shear rate).

net pressure: the pressure in the fracture at any point during a fracture treatment (and at any point along the created fracture length) minus the far-field minimum principal stress in the formation (pressure at which the fracture will close).

nonionic: electrically neutral.

overburden stress: the absolute vertical stress exerted on a formation at depth by the weight of overlying formations.

overburden: the strata of rock that lie above the formation being produced or targeted for hydraulic fracturing.

overflush: fluid pumped after (or "behind") the fracturing fluid, over and above the volume necessary to displace the surface piping and downhole tubulars.

oxidizers: a high-temperature breaker (can be used at temperatures up to 325°F), often used where persulfates are too fast-acting (pH range of 3 to 14).

pay layer: oil- or gas-producing.

permeability: a measure of the ease with which fluids can flow through a porous rock, symbol k.

persulfates: a family of breakers, including encapsulated and activated varieties, that are generally economical and find application over a wide range of temperatures (70°F to 200°F), concentrations, and pH values.

petroleum: a term referring collectively to the liquid (oil) and vapor (natural gas) forms of hydrocarbons—the particular phase being determined by the sizes of the compounds, in conjunction with pressure and temperature. (See also *hydrocarbons.*)

pH: a scale used to express the degree of acidity or alkalinity of a substance, with values from 0 to 14 (the number 7 representing neutrality, numbers below 7 indicating increased acidity, and numbers above 7 increased alkalinity).

plastic fluid: a complex, non-Newtonian fluid that requires a positive shear stress (yield stress) to initiate flow but exhibits straight-line rheological behavior. (Compare to *Newtonian fluid.*)

polymer: the basic ingredient in fracturing fluids, a petroleum-based substance in which large molecules are constructed from smaller molecules in repeating structural units.

porosity: a measure of the void space within a rock, expressed as a fraction or percentage of the bulk volume of that rock, symbol f.

pressure: the application of force on or to something by something else (e.g., the force of a 20,000 ft column of water on the bottom of a hole).

proppant: a material (e.g., naturally occurring sand or manmade ceramics) used to hold open (or "prop") a fracture so that more fluid can be produced from or injected into a well.

pseudoplastic fluid: a fluid that exhibits no yield stress but for which the slope of the rheological curve decreases with increasing shear rate. (Compare to *plastic fluid* and *Newtonian fluid.*)

radial flow: the simplest and logical converging flow pattern that results from fluids flowing into a vertical well from the surrounding drainage area.

reservoir: a porous, permeable rock formation that contains oil and/or natural gas (and always accompanied by water, whether producible or immovable) enclosed or surrounded by layers of less permeable rock.

reservoir pressure: the fluid pressure in a petroleum-bearing formation, expressed either as *initial* reservoir pressure, symbol p_i, *average* reservoir pressure, symbol $\bar{p}$, or *outer boundary constant pressure,* symbol p_e. (Shut-in bottomhole pressures measured at the formation face are sometimes reported as reservoir pressure. Rarely is this a valid indication of reservoir pressure.)

seismic imaging: detailed information obtained from the acoustic response—reflection and refraction—of various earth strata to artificial vibrations created at the earth's surface or in wells.

settling rate: vertical distance (in feet) that a particle will travel in one minute through a static fluid.

skin effect: a dimensionless term incorporated in production rate calculations to account for deviations in well performance caused by formation damage in the near-well area, symbol s.

slot flow: the laminar fluid flow regime that occurs in a channel of rectangular cross section when the ratio of fracture height to fracture length is extremely large, corresponding to the KGD geometry.

specific gravity: a ratio of the weight of a given volume of a solid or liquid to the weight of the same volume of pure water at the same temperature, used as a means of comparison. (Water, as the most plentiful matter on earth, was selected as the basis for weight comparisons and assigned the arbitrary but convenient specific gravity of 1.0. Gas specific gravity is measured and reported with respect to air at standard conditions.)

stimulation: increasing the fluid flow capacity from (or into) sandstone or carbonate formations by acidizing, fracture acidizing, or hydraulic fracturing.

stresses at depth: a system of three principal stresses, one vertical and two horizontal, to which a formation at depth is subjected. (These are also the far-field stresses.)

surface tension: the force acting within the interface between a liquid and its own vapor, which tends to maintain the surface area at a minimum. (See also *interfacial tension*.)

surfactant: a material that alters physical characteristics or properties, such as interfacial tension or wettability between fluids and solids. Surface-active agents may be classified as emulsifiers, de-emulsifiers, wetting agents, foamers, and dispersing agents.

tensile force: force placed on a rock in the opposite direction of compressive force, which creates a fracture or crack.

turbulent flow: fluid flow in which secondary irregularities and eddies are imposed on the main or average flow pattern. (See also *laminar flow*.)

vertical fracture: the most common type of hydraulic fracture, often envisioned as two symmetric wings emanating from a vertical wellbore in a single plane.

vertical stress: see *overburden stress*.

viscosity: a measure of the resistance of a fluid to flow, symbol μ. (The viscosity of petroleum is typically expressed in terms of the time required for a specific volume of liquid to flow through a calibrated opening.)

well completion: the activities and methods necessary to prepare a well for the production of petroleum, establishing a flow conduit between the reservoir and the surface.

Bibliography

Acharya, A. (1986). "Particle Transport in Viscous and Viscoelastic Fracturing Fluids," *SPEPE* (March) 104–110.

Advani, S. H. (1982). "Finite Element Model Simulations Associated with Hydraulic Fracturing," *SPEJ* (April) 209–218.

Agarwal, R. G., et al. (1979). "Evaluation and Performance Prediction of Low-Permeability Gas Wells Stimulated by Massive Hydraulic Fracturing," *JPT* (March) 362–372.

Aggour T. M. and Economides, M. J. (1996). "Impact of Fluid Selection on High-Permeability Fracturing," paper SPE 36902.

Ayoub, J. A., Barree, R. D. and Chu, W. C. (2000). "Evaluation of Frac and Pack Completions and Future Outlook," (SPE 65063) *SPEPF* (August).

Ayoub, J. A., Kirksey, J. M., Malone, B. P. and Norman, W. D. (1992). "Hydraulic Fracturing of Soft Formations in the Gulf Coast," paper SPE 23805.

Babcock, R. E., Prokop, C. L. and Kehle, R. O. (1967). "Distribution of Propping Agents in Vertical Fractures," *Producers Monthly* (November) 11–18.

Balen, R. M., Meng, H. Z. and Economides, M. J. (1988). "Application of the Net Present Value (NPV) in the Optimization of Hydraulic Fractures," paper SPE 18541.

Baree, R. D., Rogers, B. A. and Chu, W. C. (1996). "Use of Frac-Pac Pressure Data to Determine Breakdown Conditions and Reservoir Properties," paper SPE 36423.

Barree, R. D. (1983). "A Practical Numerical Simulator for Three Dimensional Fracture Propagation in Heterogeneous Media," paper SPE 12273.

Barree, R. D. and Conway, M. W. (1995). "Experimental and Numerical Modeling of Convective Proppant Transport," *JPT* (March) 216.

Barree, R. D. and Mukherjee, H. (1995). "Engineering Criteria for Flowback Procedures," paper SPE 29600.

Boutéca, M. J. (1988). "Hydraulic Fracturing Model Based on a Three-Dimensional Closed Form: Tests and Analysis of Fracture Geometry and Containment," *SPEPE* (November) 445–454, *Trans. AIME* **285**.

Britt, L. K. (1985). "Optimized Oilwell Fracturing of Moderate Permeability Reservoirs," paper SPE 14371.

Brown, J. E. and Economides, M. J. (1992). "An Analysis of Hydraulically Fractured Horizontal Wells," paper SPE 24322.

Brown, J. E., King. L. R., Nelson, E. B. and Ali, S. A. (1996). "Use of a Viscoelastic Carrier Fluid in Frac-Pack Applications," paper SPE 31114.

Castillo, J. L. (1987). "Modified Pressure Decline Analysis Including Pressure-Dependent Leak-off," paper SPE 16417.

Chambers, D. J. (1994). "Foams for Well Stimulation in Foams: Fundamentals and Applications in the Petroleum Industry," *ACS Advances in Chem. Ser.* **242**, 355–404.

Chapman, B. J., Vitthal, S. and Hill, L. M. (1996). "Pre-fracturing Pump-in Testing for High Permeability Formations," paper SPE 31150.

Chudnovsky, A., Fan, J., Shulkin, Y., Dudley, J. W., Shlyapobersky, J. and Schraufnagel, R. (1996). "A New Hydraulic Fracture Tip Mechanism in a Statistically Homogeneous Medium," paper SPE 36442.

Cikes, M. (2000). "Long-Term Hydraulic Fracture Conductivities Under Extreme Conditions," (SPE 66549) *SPEPF* (November).

Cinco, H. L., et al. (1981). "Transient Pressure Analysis for Fractured Wells," *JPT* (September) 1749–1766.

Cinco-Ley, H., Samaniego, F., Dominguez, F. (1978). "Transient Pressure Behavior for a Well with Finite-Conductivity Vertical Fracture," *SPEJ*, 253–264.

Cinco-Ley, H. and Samaniego, F. (1981). "Transient Pressure Analysis for Fractured Wells," *JPT*, 1749–1766.

Cinco-Ley, H. and Samaniego-V., F. (1981). "Transient Pressure Analysis: Finite Conductivity Fracture Case Versus Damage Fracture Case," paper SPE 10179.

Cleary, M. P. and Fonseca, A., Jr. (1992). "Proppant Convection and Encapsulation in Hydraulic Fracturing: Practical Implications of Computer and Laboratory Simulations," paper SPE 24825.

Clifton, R. J. and Abou-Sayed, A. S. (1979). "On the Computation of the Three-Dimensional Geometry of Hydraulic Fractures," paper SPE 7943.

Conway, M. W., McGowen, J. M., Gunderson, D. W. and King, D. (1985). "Prediction of Formation Response from Fracture Pressure Behavior," paper SPE 14263.

Daneshy, A. A. (1978): "Numerical Solution of Sand Transport in Hydraulic Fracturing," *JPT* (November) 132–140.

de Pater, C. J., Weijers, L., Savic, M., Wolf, K-H. A. A., van den Hoek, P. J. and Barr, D. T. (1993). "Experimental Study of Nonlinear Effects in Hydraulic Fracture Propagation," paper SPE 25893.

DeBonis, V. M., Rudolph, D. A. and Kennedy, R. D. (1994). "Experiences Gained in the Use of Frac-Packs in Ultra-Low BHP Wells, U.S. Gulf of Mexico," paper SPE 27379.

Detournay, E. and Cheng, A. H-D. (1988). "Poroelastic Response of a Borehole in a Non-hydrostatic Stress Field," *Int. J. Rock Mech., Min. Sci. and Geomech. Abstr.* **25** (3) 171–182.

Dusterhoft, R., Vitthal, S., McMechan, D. and Walters, H. (1995). "Improved Minifrac Analysis Technique in High-Permeability Formations," paper SPE 30103.

Ebinger, C. D. (1996). "New Frac-Pack Procedures Reduce Completion Costs," *World Oil* (April).

Economides, M. and Nolte, K. (Eds.) (1989). *Reservoir Stimulation* (2nd ed.), Prentice Hall, Englewood Cliffs, NJ.

Ehlig-Economides, C. A., Fan Y. and Economides, M. J. (1994). "Interpretation Model for Fracture Calibration Tests in Naturally Fractured Reservoirs," paper SPE 28690.

Elbel, J. L., et al. (1987). "Use of Cumulative-Production Type Curves in Fracture Design," *SPEPE* (August) 191–198.

Elbel, J. L., Navarrete, R. C. and Poe, B. D., Jr. (1995). "Production Effects of Fluid Loss in Fracturing High-Permeability Formations," paper SPE 30098.

Ely, J. W. (1994). *Stimulation Engineering Handbook,* Pennwell, Houston.

Fan, Y. and Economides, M. J. (1995). "Fracturing Fluid Leakoff and Net Pressure Pressure Behavior in Frac&Pack Stimulation," paper SPE 29988.

Firoozabadi, A. and Katz, D. L. (1979). "An Analysis of High-Velocity Gas Flow Through Porous Media," *JPT* (February) 211–216.

Fisher, K. (2001). "Fracture Diagnostics Case Histories," *GasTIPS* **7** (3) 9–14.

Frederick D. C., Jr., et al. (1994). "New Correlations to Predict Non-Darcy Flow Coefficients at Immobile and Mobile Water Saturation," paper SPE 28451.

Geertsma, J. and de Klerk, F. (1969). "A Rapid Method of Predicting Width and Extent of Hydraulically Induced Fractures," *JPT* (December) 1571–1581.

Gidley, J. L., Holditch, S. A., Nierode, D. E. and Veatch, R.W., Jr. (Eds.) (1989). *Recent Advances in Hydraulic Fracturing*, Monograph 12, SPE, Richardson, TX.

Gidley, J. L. (1990). "A Method for Correcting Dimensionless Fracture Conductivity for Non-Darcy Flow Effects," paper SPE 20710.

Gringarten, A. C. and Ramey, A. J., Jr. (1974). "Unsteady State Pressure Distributions Created by a Well with a Single-Infinite Conductivity Vertical Fracture," *SPEJ* (August) 347–360.

Grubert, D. M. (1990). "Evolution of a Hybrid Frac-Gravel Pack Completion: Monopod Platform, Trading Bay Field, Cook Inlet, Alaska," paper SPE 19401.

Guppy, K. H, Cinco-Ley, H., Ramey, H. J., Jr. and Samaniego-V., F. (1982). "Non-Darcy Flow in Wells With Finite-Conductivity Vertical Fractures," *SPEJ* (April) 681–698, *Trans. AIME* **273**.

Haimson, B. C. and Fairhurst, C. (1967). "Initiation and Extension of Hydraulic Fractures in Rocks," *SPEJ* (September) 310–318.

Hannah, R. R., Park, E. I., Walsh, R. E., Porter, D. A., Black, J. W. and Waters, F. (1993). "A Field Study of a Combination Fracturing/Gravel Packing Completion Technique on the Amberjack, Mississippi Canyon 109 Field," paper SPE 26562.

Holditch, S. A. and Morse, R. A. (1976). "The Effects of Non-Darcy Flow on the Behavior of Hydraulically Fractured Gas Wells," *JPT* (October) 1169–1178.

Holditch, S. A. and Morse, R. A. (1976). "The Effects of Non Darcy Flow on the Behavior of Hydraulically Fractured Gas Wells," *JPT* (October) 1159–1178.

Howard, G. C. and Fast, C. R. (1957): "Optimum Fluid Characteristics for Fracture Extension," *API Drilling and Production Prac. API*, 261–270.

Hubbert, M. K. and Willis, D. G. (1957). "Mechanics of Hydraulic Fracturing," *Trans. AIME* **210**, 153–166.

Hunt, J. L. and Soliman, M. Y. (1994). "Reservoir Engineering Aspects of Fracturing High-Permeability Formations," paper SPE 28803.

Hunt, J. L., Chen, C.-C. and Soliman, M. Y. (1994). "Performance of Hydraulic Fractures in High-Permeability Formations," paper SPE 28530.

Jin, L. and Penny, G. S. (2000). "A Study of Two-Phase, Non-Darcy Gas Flow Through Proppant Pacs," (SPE 66544) *SPEPF* (November).

Keck, R. G., Hainey, B. W. and Clausen, R. A. (1993). "An Integrated Laboratory and Field Study to Optimize the Hydraulic Fracturing Fluid System in High-Permeability," paper SPE 25467.

Khristianovitch, S. A. and Zheltov, Y. P. (1955). "Formation of Vertical Fractures by Means of Highly Viscous Fluids," *Proc., World Pet. Cong., Rome* **2**, 579–586.

Kirby, R. L., Clement, C. C., Asbill, S. W., Shirley, R. M. and Ely, J. W. (1995). "Screenless Frac Pack Completions Utilizing Resin Coated Sand in the Gulf of Mexico," paper SPE 30467.

Lacy, L. L., Rickards, A. R. and Ali, S. A. (1997). "Embedment and Fracture Conductivity in Soft Formations Associated with HEC, Borate and Water-Based Fracture Designs," paper SPE 38590.

Ledlow, L. B., Johnson, M. H., Richard, B. M. and Huval, T. J. (1993). "High-Pressure Packing with Water: An Alternative Approach to Conventional Gravel Packing," paper SPE 26543.

Martins, J. P., Collins, P. J. and Rylance, M. (1992). "Small Highly Conductive Fractures Near Reservoir Fluid Contacts: Application to Prudhoe Bay," paper SPE 24856.

Martins, J. P., Leung, K. H., Jackson, M. R., Stewart, D. R. and Carr, A. H. (1989). "Tip Screenout Fracturing Applied to the Ravenspurn South Gas Field Development," paper SPE 19766.

Mathur, A. K., Ning, X., Marcinew, R.B., Ehlig-Economides, C. A. and Economides, M. J. (1995). "Hydraulic Fracture Stimulation of High-Permeability Formations:

The Effect of Critical Fracture Parameters on Oilwell Production and Pressure," paper SPE 30652.

Mayerhofer, M. J., Economides, M. J. and Ehlig-Economides, C. A. (1995). "Pressure-Transient Analysis of Fracture-Calibration Tests," *JPT* (March) 1–6.

Mayerhofer, M. J., Economides, M. J. and Nolte, K. G. (1991): "An Experimental and Fundamental Interpretation of Fracturing Filter-Cake Fluid Loss," paper SPE 22873.

McGowen, J. M. and Vitthal S. (1995). "Fracturing Fluid Leakoff Under Dynamic Conditions Part 1: Development of a Realistic Laboratory Testing Procedure," paper SPE 36492.

McGowen, J. M., Vitthal, S., Parker, M. A., Rahimi, A. and Martch, W. E., Jr. (1993). "Fluid Selection for Fracturing High-Permeability Formations," paper SPE 26559.

McGuire, W. J. and Sikora, V. J. (1960). "The Effect of Vertical Fractures on Well Productivity," *JPT* (October) 72.

McLarty, J. M. and DeBonis, V. (1995). "Gulf Coast Section SPE Production Operations Study Group—Technical Highlights from a Series of Frac Pack Treatments," paper SPE 30471.

Mears, R. B., Sluss, J. J., Jr., Fagan, J. E. and Menon, R. K. (1993). "The Use of Laser Doppler Velocimetry (LDV) for the Measurement of Fracturing Fluid Flow in the FFCF Simulator," paper SPE 26619.

Medlin, W. L. and Fitch, J. L. (1988). "Abnormal Treating Pressures in Massive Hydraulic Fracturing Treatments," *JPT* 633–642.

Meng, H. Z. (1987). "Coupling of Production Forecasting, Fracture Geometry Requirements and Treatment Scheduling in the Optimum Fracture Design," paper SPE/DOE 16435.

Meyer, B. R and Hagel, M. W. (1989). "Simulated Mini-Fracs Analysis," *JCPT* **28** (5) 63–73.

Milton-Tayler, D. (1993). "Non-Darcy Gas Flow: From Laboratory Data to Field Prediction," paper SPE 26146.

Montagna, J. N., Saucier, R. J. and Kelly, P. (1995). "An Innovative Technique for Damage By-pass in Gravel Packed Completions Using Tip Screenout Fracture Prepacks," paper SPE 30102.

Montgomery, K. T., et al. (1990). "Effects of Fracture Fluid Invasion on Cleanup Behavior and Pressure Buildup Analysis," paper SPE 20643.

Monus, F. L., Broussard, F. W., Ayoub, J. A. and Norman, W. D. (1992). "Fracturing Unconsolidated Sand Formations Offshore Gulf Mexico," paper SPE 22844.

Morales, R. H. and Marcinew, R. P. (1993): "Fracturing of High-Permeability Formations: Mechanical Properties Correlations," paper SPE 26561.

Morse, R. A. and Von Gonten, W. D. (1971). "Productivity of Vertically Fractured Wells Prior to Stabilized Flow," paper SPE 3631.

Mullen, M. E., Norman, W. D. and Granger, J. C. (1994). "Productivity Comparison of Sand Control Techniques Used for Completions in the Vermilion 331 Field," paper SPE 27361.

Mullen, M. E., Norman, W. D., Wine, J. D. and Stewart, B. R. (1996). "Investigation of Height Growth in Frac Pack Completions," paper SPE 36458.

Mullen, M. E., Stewart, B. R. and Norman, W. D. (1994). "Evaluation of Bottomhole Pressures in 40 Soft Rock Frac-Pack Completions in the Gulf of Mexico," paper SPE 28532.

Nierode, D. E. and Kruk, K. F. (1973). "An Evaluation of Acid Fluid Loss Additives Retarded Acids, and Acidized Fracture Conductivity," paper SPE 4549.

Ning, X., Marcinew, R. P. and Olsen, T. N. (1995). "The Impact of Fracturing Fluid Cleanup and Fracture-Face Damage on Gas Production," paper CIM 95-43.

Nolte, K. G. (1986). "Determination of Proppant and Fluid Schedules from Fracturing Pressure Decline," (SPE 8341) *SPEPE* (July) 255–265.

Nolte, K. G. and Smith, M. B. (1981). "Interpretation of Fracturing Pressures," *JPT* (September) 1767–1775.

Nolte, K. G., Mack, M. G. and Lie W. L. (1993). "A Systematic Method for Applying Fracturing Pressure Decline: Part 1," paper SPE 25845.

Nordgren, R. P. (1972). "Propagation of a Vertical Hydraulic Fracture," *SPEJ* (August) 306–314, *Trans. AIME* **253.**

Novotny, E. J. (1977). "Proppant Transport," paper SPE 6813.

Palmer, I. D. and Veatch, R. W., Jr. (1990). "Abnormally High Fracturing Pressures in Step-Rate Tests," *SPEPE* (August) 315–323, *Trans. AIME* **289.**

Parker, M. A., Vitthal, S., Rahimi, A., McGowen, J. M., and Martch, W. E., Jr. (1994). "Hydraulic Fracturing of High-Permeability Formations to Overcome Damage," paper SPE 27378.

Parlar, M., Nelson, E. B., Walton, I. C., Park, E. and DeBonis, V. (1995). "An Experimental Study on Fluid-Loss Behavior of Fracturing Fluids and Formation Damage in High-Permeability Porous Media," paper SPE 30458.

Patel, Y. K., Troncoso, J. C., Saucier, R. J. and Credeur, D. J. (1994). "High-Rate Pre-Packing Using Non-Viscous Carrier Fluid Results in Higher Production Rates in South Pass Block 61 Field," paper SPE 28531.

Penny, G. S., et al. (1987). "An Evaluation of the Effects of Environmental Conditions and Fracturing Fluids Upon the Long Term Conductivity of Proppants," paper SPE 16900.

Penny, G. S., et al. (1996). "The Use of Inertial Force and Low Shear Viscosity to Predict Cleanup of Fracturing Fluids within Proppant Packs," paper SPE 31096.

Perkins, T. K. and Kern, L. R. (1961). "Width of Hydraulic Fractures," *JPT* (September) 937–949, *Trans. AIME* **222.**

Prats, M. (1961). "Effect of Vertical Fractures on Reservoir Behavior—Incompressible Fluid Case," *SPEJ* (June) 105–118, *Trans. AIME* **222.**

Rahim, Z. and Holditch, S. A. (1993). "Using a Three-Dimensional Concept in a Two-Dimensional Model to Predict Accurate Hydraulic Fracture Dimensions," paper SPE 26926.

Reidenbach, V. G., Harris, P. C., Lee, Y. N. and Lord, D. L. (1986). "Rheological Study of Foam Fracturing Fluids Using Nitrogen and Carbon Dioxide," *SPEPE* (January) 39–41.

Reimers, D. R. and Clausen, R. A. (1991). "High-Permeability Fracturing at Prudhoe Bay, Alaska," paper SPE 22835.

Robinson, B. M., et al. (1988). "Minimizing Damage to a Propped Fracture by Controlled Flowback Procedures," *JPT* (June) 753–759.

Roodhart, L. P. (1985). "Proppant Settling in Non-Newtonian Fracturing Fluids," paper SPE 13905.

Roodhart, L. P., Fokker, P. A., Davies, D. R., Shlyapobersky, J. and Wong, G. K. (1993). "Frac and Pack Stimulation: Application, Design, and Field Experience from the Gulf of Mexico to Borneo," paper SPE 26564.

Roodhart, L. P. (1985). "Fracturing Fluids: Fluid-Loss Measurements Under Dynamic Conditions," *SPEJ* (October) 629–636.

Saucier, R. J. (1974). "Considerations in Gravel Pack Design," *JPT* (February) 205–212.

Settari, A. (1985). "A New General Model of Fluid Loss in Hydraulic Fracturing," *SPEJ*, 491–501.

Settari, A. and Cleary, M. P. (1986). "Development and Testing of a Pseudo-Three-Dimensional Model of Hydraulic Fracture Geometry," *SPEPE* (November) 449–466, *Trans. AIME* **283.**

Shah, S. N. (1982). "Proppant Settling Correlations for Non-Newtonian Fluids Under Static and Dynamic Conditions," *SPEJ*, 164–170.

Shlyapobersky J., Walhaug, W. W., Sheffield. R. E. and Huckabee, P. T. (1988). "Field Determination of Fracturing Parameters for Overpressure Calibrated Design of Hydraulic Fracturing," paper SPE 18195.

Shlyapobersky, J., Wong, G. K. and Walhaugh, W. W. (1988). "Overpressure-Calibrated Design of Hydraulic Fracture Simulations," paper SPE 18194.

Simonson, E. R., Abou-Sayed, A. S. and Clifton, R. J. (1978). "Containment of Massive Hydraulic Fractures," *SPEJ,* 27–32.

Singh, P. K. and Agarwal, R. G. (1988). "Two-Step Rate Test: A New Procedure for Determining Formation Parting Pressure," paper SPE 18141.

Smith, M. B. and Hannah, R. R. (1994). "High-Permeability Fracturing: The Evolution of a Technology," paper SPE 27984.

Smith, M. B., Miller, W. K., II and Haga, J. (1987). "Tip Screenout Fracturing: A Technique for Soft, Unstable Formation," *SPEPE* (May) 95–103.

Sneddon, I. N. (1973). "Integral Transform Methods," chapter in *Mechanics of Fracture I—Methods of Analysis and Solutions of Crack Problems*, G. C. Sih (Ed.), Nordhoff International, Leyden.

Soliman, M. Y. (1986). "Technique for Considering Fluid Compressibility Temperature Changes in Mini-Frac Analysis," paper SPE 15370.

Soliman, M. Y., et al. (1985). "Effect of Fracturing Fluid and Its Cleanup on Well Performance," paper SPE 14514.

Soliman, M. Y. and Daneshy, A. A. (1991). "Determination of Fracture Volume and Closure Pressure from Pump-In/Flowback Tests," paper SPE 21400.

Stewart, B. R., Mullen, M. E., Ellis, R. C., Norman, W. D. and Miller, W. K. (1995). "Economic Justification for Fracturing Moderate to High-Permeability Formations in Sand Control Environments," paper SPE 30470.

Stewart, B. R., Mullen, M. E., Howard, W. J. and Norman, W. D. (1995). "Use of a Solids-Free Viscous Carrying Fluid in Fracturing Applications: An Economic and Productivity Comparison in Shallow Completions," paper SPE 30114.

Thiercelin, M. J, Ben-Naceur, K. and Lemanczyk, Z. R. (1985). "Simulation of Three-Dimensional Propagation of a Vertical Hydraulic Fracture," paper SPE 13861.

Tiner, R. L., Ely, J. W. and Schraufnagel, R. (1996). Frac Packs—State of the Art," paper SPE 36456.

Tinker, S. J., Baycroft, P. D., Elis, R. C. and Fitzhugh, E. (1997). Mini-Frac Tests and Bottomhole Treating Pressure Analysis Improve Design and Execution of Fracture Stimulatons," paper SPE 37431.

Valkó, P. and Economides, M. J. (1995). *Hydraulic Fracture Mechanics*, Wiley, Chichester.

Valkó, P., Oligney, R. E. and Schraufnagel, R. A. (1996). "Slopes Analysis of Frac & Pack Bottomhole Treating Pressures," paper SPE 31116.

Valkó, P. and Economides, M. J. (1996). "Performance of Fractured Horizontal Wells in High-Permeability Reservoirs," paper SPE 31149.

van Poollen, H. K., Tinsley, J. M. and Saunders, C. D. (1958). "Hydraulic Fracturing—Fracture Flow Capacity vs. Well Productivity," *Trans. AIME* **213**, 91.

Vinegar, H. J., Willis, P. B., DeMartini, D. C., Shlyapobersky, J., Deeg, W. F. J., Adair, R. G., Woerpel, J. C., Fix, J. E. and Sorrells, G. G. (1992). "Active and Passive Seismic Imaging of a Hydraulic Fracture in Diatomite," *JPT* (January) 28–34, 88–90.

Vitthal S. and McGowen J. M. (1995). "Fracturing Fluid Leakoff Under Dynamic Conditions Part 2: Effect of Shear Rate, Permeability and Pressure," paper SPE 36493.

Warenbourg, P. A., et al. (1985). "Fracture Stimulation Design and Evaluation," paper SPE 14379.

Williams, B. B. (1970). "Fluid Loss From Hydraulically Induced Fractures," *JPT*, 882–888, *Trans. AIME* **249**.

Winkler, W., Valkó, P. and Economides, M. J. (1994). "Laminar and Drag-Reduced Polymeric Foam Flow," *J. of Rheology* **38**, 111–127.

Wong, G. K., Fors, R. R., Casassa, J. S., Hite, R. H. and Shlyapobersky, J. (1993). "Design, Execution and Evaluation of Frac and Pack (F&P) Treatments in Unconsolidated Sand Formations in the Gulf of Mexico," paper SPE 26564.

Wong, S. W. (1970). "Effect of Liquid Saturation on Turbulence Factors for Gas-Liquid Systems," *J. of Canadian Pet. Tech.* (October–December) 274.

Yadavalli, S. K. and Jones, J. R. (1996). "Interpretation of Pressure Transient Data from Hydraulically Fractured Gas Condensate Wells," paper SPE 36556.

Yi, T. and Peden, J. M. (1994). "A Comprehensive Model of Fluid loss in Hydraulic Fracturing," *SPEPF* (November) 267–272.

Fracture Design Spreadsheet

The HF2D Excel spreadsheet is a fast 2D design package for traditional (moderate permeability and hard rock) and frac & pack (higher permeability and soft rock) fracture treatments.

It contains the following worksheets:

- Traditional design with the PKN (Perkins-Kern-Nordgren) model

- TSO (tip screenout) design with the PKN model

- CDM (Continuum Damage Mechanics) design with the PKN model

The uniqueness of this package comes from the design logic on which it is based. The design starts from the available mass of proppant; then optimum fracture dimensions are determined; finally, a treatment schedule is calculated to achieve optimum placement of the proppant. If physical constraints prohibit optimum placement, a sub-optimal placement is recommended.

Information related to the design spreadsheet included elsewhere in the text is not repeated here, for example assumptions about theoretical fracture performance (Chapter 3), the suggested design procedure based on optimal pseudo-steady state performance (Chapter 7) and sample runs (Chapter 8). After reviewing this appendix, the reader is encouraged to work stepwise through the examples given in Chapter 8 to gain a quick working sense of the spreadsheet.

As it is, the appendix is limited to a specific description of input data required to run the spreadsheet, and calculated results. The results include fluid and proppant requirements, injection rates, added proppant concentrations (i.e., the proppant schedule), and additional information on the evolution of fracture dimensions.

DATA REQUIREMENT

The following table contains a description of the required input parameters.

Input Parameter	Remark
Proppant mass for (two wings), lb$_m$	Proppant volume (mass) is the single most important decision variable of the design procedure.
Sp grav of proppant material (water = 1)	For instance, 2.65 for sand.
Porosity of proppant pack	The porosity of the pack might vary with closure stress; a typical value is 0.3.
Proppant pack permeability, md	Retained permeability, including fluid residue and closure stress effects, might be reduced by a factor as large as 10 in case of non-Darcy flow in the fracture. Realistic proppant pack permeability is in the range of 10,000 to 100,000 md for in-situ flow conditions. Values provided by manufacturers, such as 500,000 md for a "high strength" proppant, should be considered with caution.
Max propp diameter, D_{pmax}, inch	From mesh size; for 20/40 mesh sand it is 0.035 in.
Formation permeability, md	Effective permeability of the formation.
Permeable (leakoff) thickness, ft	This parameter is used in calculation of the Productivity Index (as net thickness) and the apparent leakoff coefficient—assuming no leakoff (or spurt loss) outside of the permeable thickness.
Well radius, ft	Needed for pseudo-skin factor calculation.
Well drainage radius, ft	Needed for optimum design. (Do not underestimate the importance of this parameter!)

Input Parameter (cont.)	Remark
Pre-treatment skin factor	Can be set to zero as it does not influence the design. Is only used as a basis for calculating the "folds of increase" in productivity.
Fracture height, ft	Usually greater than the permeable height. One of the most critical design parameters. Derived from lithology information, or can be adjusted iteratively by the user to roughly match the fracture length.
Plane strain modulus, E' (psi)	Defined as Young's modulus divided by one minus the Poisson ratio squared, $E' = E/(1 - n^2)$. Is almost the same as Young's modulus, and about twice the shear modulus (the effect of the Poisson ratio is small). For hard rock, it might be 10^6 psi, for soft rock 10^5 psi or less.
Slurry injection rate (two wings, liq + prop), bpm	The injection rate is considered constant. It includes both the fracturing fluid and the proppant. Additional proppant simply reduces the calculated liquid injection rate. Typical value is 30 bpm.
Rheology, K' $(lb_f/ft^2)*s^{n'}$	Power law consistency of the fracturing fluid (slurry, in fact).
Rheology, n'	Power law flow behavior index.
Leakoff coefficient in permeable layer, $ft/min^{1/2}$	Leakoff outside the permeable layer is considered zero, so when the ratio of fracture height to permeable-layer thickness is high, the apparent leakoff coefficient calculated from this input is much lower than the input itself. If leakoff is suspected outside of the net pay, this parameter may be adjusted along with fracture height.
Spurt loss coefficient, S_p, gal/ft²	Accounts for spurt loss in the permeable layer. Outside of the permeable layer, spurt loss is considered zero (see previous remark).
Max possible added proppant concentration, lb_m/gallon neat fluid	The most important equipment constraint. Some current mixers can provide more than 15 lb_m/gal of neat fluid. Often it is not necessary to ramp up to the maximum possible concentration.
Multiply opt length by factor	This design parameter can be used to force a sub-optimal design. If the optimum length is considered too small (fracture width too large), a value greater than one is used. If the optimum length is too large (fracture width too

Input Parameter (cont.)	Remark
	small), a fractional value may be used. This possibility of user intervention is handy for investigating the pros and cons of departing from the technical optimum. The default value is 1. See more on this issue in the text of Chapter 8.
Multiply Nolte pad by factor	In accordance with Nolte's suggestion, the exponent of the proppant concentration schedule and the pad fraction (relative to total injected volume) are initially taken to be equal (i.e., represented by the default input value of 1). Inputting a value other than 1 has the effect of increasing or decreasing the pad fraction accordingly. The program adjusts the proppant schedule to ensure the required amount of proppant is injected.

Additional input parameters are required for the special cases of TSO and CDM designs.

Input Parameter	Remark
TSO criterion Wdry/Wwet	Specifies the ratio of dry width (when only dehydrated proppant is left in the fracture) to wet width (dynamically achieved during pumping), and is needed only for the TSO design. According to our assumption, screenout happens when the ratio of dry-to-wet width reaches the user specified value. We suggest using a number between 0.5 and 0.75 initially, but this number should be refined locally based on evaluation of successful TSO treatments.
Closure pressure, psi	Closure pressure must be specified for the PKN-CDM design, whereas the traditional PKN and PKN-TSO designs are based on net pressures. For example, the Continuum Damage Mechanics model of fracture propagation velocity is affected by the absolute value of the minimum stress.
CDM Cl^2, ft^2/(psi-sec)	This combined CDM parameter (together with closure pressure) influences the fracture propagation velocity. If this value is large (for instance, on the order of 1), fracture

Input Parameter (cont.)	Remark
	propagation is not retarded; essentially, the model behaves as a traditional PKN model. When the value is relatively small, such as 0.01 ft^2/(psi-sec), fracture propagation is retarded. It takes more time to reach a given length, and the width and net pressure are higher than calculated with the traditional PKN model. This design parameter can be estimated by matching the excess net pressure experienced during the minifrac treatment.

CALCULATED RESULTS

The results comprise *theoretical optimum* and *actual* placement details. One may or may not be able to achieve the technical optimum fracture dimensions, depending on certain constraints (e.g., maximum proppant concentration). A boldface red message appears in the spreadsheet to denote when optimum fracture dimensions cannot be achieved.

The main fracture characteristics, such as half-length, average width and areal proppant concentration, determine the performance of the fractured well. Fracture performance is given in terms of dimensionless productivity index and as a post-treatment pseudo-skin factor.

Cumulative fluid and proppant volumes are reported. Fluid injection rate and the added proppant concentration are presented as functions of time.

The results include:

Output Parameter	Remark
t, min	Time elapsed from start of pumping
qi_liq, bpm	Liquid injection rate (for two wings)
cum liq, gal	Cumulative liquid injected up to time t
cadd, lb$_m$/gal	Proppant added to one gallon of liquid, in other words ppga
cum prop, lb$_m$	Cumulative proppant injected up to time t
x_f, ft	Half-length of the fracture at time t
w_{ave}, in.	Average width of the fracture at time t
w_{ave} / D_{pmx}	Ratio of average width of the fracture to the maximum proppant diameter, should be at least 3

Output Parameter (cont.)	Remark
w_{dry} / w_{wet}	Ratio of dry-to-wet width. During pumping, the wet width is 2 to 10 times larger than the dry width that would be necessary to contain the same amount of proppant (without any fluid and packed densely). Should be maintained at a relatively low value (e.g., 0.2) to avoid screenout during the job. Tip screenout in the TSO version of the design spreadsheet is formulated in terms of this variable.

Minifrac
Spreadsheet

The MF Excel spreadsheet is a minifrac (calibration test) evaluation package. Its main purpose is to extract the leakoff coefficient from pressure fall-off data.

It contains the following worksheets:

- Analysis with the PKN model, Nolte-Shlyapobersky method.

- Analysis with the KGD model, Nolte-Shlyapobersky method.

- Analysis with the Radial model, Nolte-Shlyapobersky method.

- Analysis with the PKN-CDM model, which includes an estimate of fracture propagation retardation (in the form of the Continuum Damage parameter, Cl^2).

The PKN-CDM option uses the "overpressure" observed during a minifrac treatment to estimate the deviation in fracture dimensions from the traditional PKN model. There is no CDM option for the KGD and Radial models.

The basic result of the analysis is the *apparent leakoff coefficient.* The apparent leakoff coefficient describes the leakoff with respect to the total created fracture area. From this *apparent* leakoff coefficient, then, a "true" leakoff coefficient is calculated—which value is valid only for the permeable layer and assumes there is no leakoff outside the permeable layer.

In case of the CDM analysis, an additional parameter is determined from the "overpressure" (i.e., the additional net pressure arising during the minifrac that cannot be explained by the traditional PKN model). The additional parameter is the CDM parameter: Cl^2, measured in ft/(psi-sec). The obtained value may be used as an input parameter in the CDM version of the FD Excel design spreadsheet.

The theory of minifrac analysis—more particularly pressure decline analysis—is included in Chapter 7, and is not repeated here. Rather, this appendix is limited to a specific description of data required to run the MF Excel spreadsheet and the results that are calculated. A sample run is also provided.

DATA REQUIREMENT

The following table contains the description of the basic input parameters.

Input Parameter	Remark
Permeable (leakoff) thickness, ft	It is assumed there is no leakoff outside the permeable thickness.
Fracture height, ft	Needed for the traditional PKN and KGD analysis, and for the PKN-CDM analysis. Is not needed for the Radial analysis.
Plane strain modulus, E' (psi)	Defined as Young modulus divided by one minus the Poisson ratio squared. Its magnitude is similar to the Young's modulus and about twice that of the shear modulus.
Closure pressure, psi	Affects the leakoff coefficient in case of the KGD and Radial models, but not in the PKN case. Also affects the CDM parameter.
Rheology, K' $(lb_f/ft^2)*s^{n'}$	Power law consistency index. Needed only for the CDM analysis.
Rheology, n'	Power law flow behavior index. Needed only for the CDM analysis

The following tabular input data are needed:

Input Parameter	Remark
t, min	Time elapsed from start of pumping
qi_liq, bpm	Liquid injection rate (for two wings, valid at the bottom of the well at the given time).
Bottomhole pressure, psi	Required only for the shut-in period, but can also be provided for the injection period. The last value during the injection period is used for "overpressure" calibration of the CDM parameter.
Include into inj vol	Flag indicator should be set to 1 for all data points (during injection period) to be included in the injection volume. Flag is set to 0 during the shut-in period.
Include into g-func fit	Flag indicator should be 1 for all data points to be included in the straight-line fit (least squares objective function) of shut-in pressure decline data. Note: Data points recorded after fracture closure are not automatically included in the fit, even if the flag is set to 1.

RESULTS

The most important result is the leakoff coefficient. The *apparent* leakoff coefficient is obtained without distinguishing between permeable and non-permeable layers. The "true" leakoff coefficient is obtained by attributing all leakoff to the permeable layer only.

As a by-product, we obtain the dimensions of the created fracture: x_f, (or R_f) and w (i.e., fracture extent and average width). The fluid efficiency is also obtained (assuming that spurt loss can be neglected). The results include:

Output Parameter	Remark
Apparent leakoff coefficient (for total area), ft/min$^{0.5}$	Can be viewed as a weighted average of the positive leakoff coefficient *inside* the permeable layer and the zero coefficient *outside*.
Leakoff coefficient in permeable layer, ft/min$^{0.5}$	A positive value. Leakoff coefficient outside the permeable layer is considered zero.
Half length, ft	Fracture half length, for PKN and KGD models.

Output Parameter (cont)	Remark
Radius, ft	Fracture radius, for the Radial model.
Efficiency (fraction)	Fracture fluid efficiency, for all models.
CDM Cl², ft²/(psi-sec)	Combined CDM parameter. Together with closure pressure, it influences fracture propagation velocity. If the value is large (for instance, on the order of 1), fracture propagation is not retarded; essentially, the model behaves as a traditional PKN model. When the value is relatively small, such as 0.01 ft²/(psi-sec), fracture propagation is retarded. It takes more time to reach a given length, and the width and net pressure are higher than calculated by the traditional PKN model. This parameter is obtained from the maximum pressure experienced during the minifrac.

SAMPLE RUN

The following example of a traditional radial minifrac analysis is provided to give the reader a quick working sense of the MF Excel spreadsheet.

Input

Permeable (leakoff) thickness, ft	42
Plane strain modulus, E' (psi)	2.00E+06
Closure Pressure, psi	5850

Tabular Input

Time from start, min	BH Injection rate, bpm	BH Pressure, psi	Include into inj volume	Include into g-func fit
0.0	9.9	0.0	1	0
1.0	9.9	0.0	1	0
2.0	9.9	0.0	1	0
3.0	9.9	0.0	1	0
4.0	9.9	0.0	1	0

Time from start, min	BH Injection rate, bpm	BH Pressure, psi	Include into inj volume	Include into g-func fit
5.0	9.9	0.0	1	0
6.0	9.9	0.0	1	0
7.0	9.9	0.0	1	0
8.0	9.9	0.0	1	0
9.0	9.9	0.0	1	0
10.0	9.9	0.0	1	0
12.0	9.9	0.0	1	0
14.0	9.9	0.0	1	0
16.0	9.9	0.0	1	0
18.0	9.9	0.0	1	0
20.0	9.9	0.0	1	0
21.0	9.9	0.0	1	0
21.5	9.9	0.0	1	0
21.8	9.9	0.0	1	0
21.95	0.0	7550.62	0	0
22.15	0.0	7330.59	0	0
22.35	0.0	7122.36	0	0
22.55	0.0	6963.21	0	1
22.75	0.0	6833.39	0	1
22.95	0.0	6711.23	0	1
23.15	0.0	6595.02	0	1
23.35	0.0	6493.47	0	1
23.55	0.0	6411.85	0	1
23.75	0.0	6347.12	0	1
23.95	0.0	6291.51	0	1
24.15	0.0	6238.43	0	1
24.35	0.0	6185.85	0	1
24.55	0.0	6135.61	0	1
24.75	0.0	6090.61	0	1
24.95	0.0	6052.06	0	1
25.15	0.0	6018.61	0	1
25.35	0.0	5987.45	0	1
25.55	0.0	5956.42	0	1

Tabular Input (cont.)

Time from start, min	BH Injection rate, bpm	BH Pressure, psi	Include into inj volume	Include into g-func fit
25.75	0.0	5925.45	0	1
25.95	0.0	5896.77	0	1
26.15	0.0	5873.54	0	1
26.35	0.0	5857.85	0	0
26.55	0.0	5849.29	0	0
26.75	0.0	5844.81	0	0
26.95	0.0	5839.97	0	0
27.15	0.0	5830.98	0	0
27.35	0.0	5816.3	0	0
27.55	0.0	5797.01	0	0
27.75	0.0	5775.67	0	0

Output

Slope, psi	−4417
Intercept, psi	13151
Injected volume, gallon	9044
Frac radius, ft	39.60
Average width, inch	0.49205
Fluid efficiency	0.16708
Apparent leakoff coefficient (for total area), ft/min$^{0.5}$	0.01592
Leakoff coefficient in permeable layer, ft/min$^{0.5}$	0.02479

Notice that the leakoff coefficient with respect to the permeable layer is greater than the apparent value, because part of the 39.6 ft radius fracture falls outside the permeable area.

Standard Practices
and QC Forms

JOB PLANNING, EXECUTION, AND POST-JOB REPORT

A traditional fracture treatment evolves in the following steps:

- An operations/production engineer at a producing company identifies a stimulation candidate(s) based on general assumptions about the type and size of the treatment and the production results that can be anticipated. (Increasingly, service company engineers are taking the initiative to identify stimulation candidates—with an immediate first step being to present a recommendation to the producing company.)

- The producing company representative meets with a stimulation/ sales engineer at a service company, conveying treatment objectives and discussing treatment specifics. A minifrac, or data frac (so named because it is designed to gather data used in optimizing the main fracture treatment), is also discussed at this time.

- The service company representative designs and recommends a minifrac and main fracture stimulation treatment. Following additional deliberation and perhaps some modifications, a pumping schedule and pricing are prepared.

239

■ The operations engineer prepares a workover program (procedure) to be followed by company field personnel in preparing the well for treatment. Such a procedure may include instructions, for example, to "kill" and clean out the well, remove the production string, and replace the production tree with a frac valve or tree saver. A sample program is provided as Appendix G.

■ A service company representative visits the well site to determine the best way to "spot" the equipment. The frac tanks are always spotted first. Not only do they consume physical space, they have an associated lead time in that they must be filled with water. A rough diagram of the location is prepared at this time.

■ The service company field manager reviews the stimulation treatment and pumping schedule step-by-step, creating a list of equipment, materials, and personnel that will be needed. (A sample checklist is provided at the end of this appendix.) Materials are sourced from the company's stock inventory or elsewhere. Next, required personnel is scheduled, taking into account maximum hours that a given driver or operator can work (under various government or company internal regulations). Consideration should be given to the fact that jobs can run longer than planned. The data frac can be pumped immediately ahead of the main fracture treatment or one day in advance, depending on the size (pumping duration) of the main treatment.

■ **Two or more days before the job:** Frac tanks are trucked to location, spotted, and filled with water (organized by either the service or producing company). Water is tested/analyzed to make sure it meets the required guidelines.

■ **One day before the job:** Small samples of the prescribed gel are mixed—preferably with water from the frac tanks—and tested. Samples of produced well fluids are tested for compatibility with the fracturing fluids and all additives that will be used during the treatment. Changes to the fluid schedule are made if necessary.

■ The service company field manager assembles the frac crew; he reviews the treatment procedure, assignments for rig-up and the main treatment, directions to location, and safety requirements.

■ **Day before/day of the job:** The Frac-Crew Chief directs the spotting of equipment on location and rig-up. He makes sure that all connections are done right, that check valves are installed properly (right side up, arrow pointing toward the well), and that all monitoring and wire control equipment is installed and functioning properly. Large treatments should be rigged-up one day in advance in order to preserve daylight hours (on the day of the treatment) for high-pressure pumping activities.

▪ **Day of the job:** The crew chief collects the pre-frac checklists, confirms the volume of chemicals and proppant on location, and gauges all tanks. Proppant sieve analysis is carried out and reported on the appropriate form. If acid is to be used, the acid quality control form is filled out, showing any dilution calculations. The supervisor is now almost ready to start the job.

- The Frac-Crew Chief seeks out the person-in-charge for the producing company.

- The QA/QC Chemist makes final tests of the water and gel samples, and fills out the Frac Fluid Blending and Quality Control form. Breaker tests are carried out and recorded on the form. Small samples are retained for the customer.

▪ **The tailgate safety meeting:** The crew chief assembles everyone on location (including his team, producing company personnel, and others) and reviews the treatment logistics:

- Assigns duties, discussing any unique testing sequence/procedures and maximum pumping pressures.

- Points out escape routes/meeting points to be used in case of emergency.

- Checks with the producing company representative to see if he has anything to add or any last minute changes.

- Sends everyone to their assigned posts.

▪ **Starting the job:** Inside the frac van, the crew chief:

- Directs final testing of all equipment and confirms that all remote monitoring equipment is functioning properly. (Each pump is pressure tested to 1,000 psi above expected treating pressure of a shut-in wellhead valve.)

- Confers with the producing company representative as to how he would like the real-time data presented, and sets up the computer accordingly.

- Obtains a final okay from the customer and commences pumping operations.

■ **During the job:** The Frac-Crew Chief keeps the customer informed of vital data such as treating pressures, stages, volumes and proppant rates, and transit times.

- The QA/QC Chemist collects samples of fracturing fluid throughout the treatment. Tests similar to those done before the job are performed and recorded. Samples of fluid and proppant are retained for the customer.

- The crew chief is prepared to make changes "on-the-fly," and offers his advice to the customer.

- The Stimulation Real Time Report form is a general guideline for items recorded before the job, and also provides a place for them to be recorded during the job. It is customary to record the ISIP and pressures at 5, 10, and 15 minutes after shutdown. The customer may request additional information (which should be determined ahead of time).

■ **After the job:** The client is informed about leftover fluid volumes, and is asked for input on how the unused fluids should be handled. A post-job summary report is provided to the producing company before departing location. All equipment is rigged down and moved, either to the next job or back to the yard. The location is thoroughly cleaned.

ADDITIONAL STANDARD PRACTICES

Mixing Acid

■ Uninhibited acid should only be stored in plastic tanks or rubber-lined steel tanks.

■ Before loading acid into a tank that is not completely empty, run compatibility tests with small samples from the tank.

■ Make sure all valves are closed.

■ Load up to 90 percent of diluting water first, and add corrosion inhibitors before loading the acid.

- Do not add gelling fluids, friction reducers, leakoff control, or corrosion intensifiers until it is time to pump.

- Show all calculations on the Acid Quality Control form.

- Always wear protective gear, such as goggles, rubber gloves, rubber apron, and respirator.

Fracturing Fluids

Always start with a visual inspection of the tanks. Gauge all tanks. Calculate both the total fluid volume and the volume of fluid available for pumping. (Depending on the design of the tanks, 10 percent of the maximum fluid volume will remain in the tanks.) Perform a water analysis and fill out the Water Quality Control form, noting any specific comments in the space provided. If biocide is required, make sure that it is added prior to loading the water.

Batch Mix Fluids

- Perform gel and crosslinker tests with all drums on location.

- Perform breaker tests.

- Prepare an exact schedule of chemicals to be added.

- Take a complete inventory of all additives, chemicals, and proppants on location.

- Gel the tanks, recording exact additive volumes.

- Perform tests on samples from each fluid tank.

- Prepare a schedule of chemicals to be added in real time.

"On-The-Fly" (Continuous) Blending

- Flush the tanks and make sure they are clean.

- Prepare an exact schedule of gelling agents, crosslinkers, and additives.

- Make sure all required chemicals and proppants are on location.

- Perform gel, crosslinker and breaker tests.

- During the job, retain small samples for testing and (potentially) inspection by the customer.

Proppants

- Take an accurate inventory of all proppants on locations.

- Perform a sieve analysis on proppant from each transport/storage compartment.

- Record results on the Proppant Sieve Analysis and Quality Control" form.

- Do not use proppant that does not meet the minimum specifications.

Safety

- Make sure that all personnel continuously wear steel-toe boots, hard-hats, and safety goggles on location.

- Equip each person that handles hazardous materials (i.e., with rubber gloves, a rubber apron, and an approved respirator). Safety equipment should be worn at all times when around the hazardous materials.

- Issue two-way radios to all personnel directly involved in the treatment. Talk should be limited to matters pertaining to the job.

- Start each job with a safety meeting, outlining procedures, personnel duties, and an emergency escape route. Designate a safe meeting place up-wind from the well location.

- Assign a vehicle and driver to be discharged on short notice in the case of an emergency. The vehicle should be parked outside the well perimeter.

QC Forms

- The following sheets are included as part of the Unified Fracture Design, both as examples and as very workable QA/QC forms that can be used by the reader:
- Stimulation Real Time Report
- Stimulation Treatment Checklist
- Fluid Blending and Quality Control
- Proppant Sieve Analysis and Quality Control

- ▪ Water Quality Control
- ▪ Acid Quality Control

UFD STIMULATION REAL TIME REPORT

DATE	CUSTOMER
FIELD	WELL

CASING AND TUBING DETAILS	PERFORATIONS		CAPACITY CALCULATIONS			
	DEPTH	SPF	ITEM	# OF FEET	BBL/FT	BBLS
			TUBING			
			CASING			
			ANNULAR			
			OPEN HOLE			
			SURFACE			
					TOTAL	

PRE-FRAC CHECKLIST

Review pump schedule with customer			Test Pumps	
All additives on location			Ask customer for final instructions	
Have safety meeting		**UFD**	Open well valve	
Determine maximum pressure			Start pumping	
Test lines to 1000# above maximum				
Check communications equipment				

STAGE	FLUID TYPE	VOLUME	RATE	(PSI)	COMMENTS
		AVERAGE			

POST JOB	
Total water pumped	
Total acid pumped	
Total gelled water pumped	
Total proppant pumped	

Pre-job SIP	
ISIP	
5 MIN	
10 MIN	
15 MIN	

$$FG = \frac{(ISIP + HH)}{TVD}$$

PLANNED ADDITIVES

ITEM	Planned # or conc.	Have on loc.	Used		ITEM	Planned # or conc.	Have on loc.	Used

UFD	STIMULATION TREATMENT CHECK LIST				

DATE			CUSTOMER		
FIELD			WELL		

ITEM	AMOUNT	UNITS	RESPONSIBILITY	YD P	LOC P
Maximum rate		BPM			
Surface Treating Pressure		PSI			
Hydraulic power (Rate*STP/40.8)		HHP			
Volume of water for data frac		BBL			
Volume of water for main frac		BBL			
Total volume of water		BBL			
No. of frac tanks (Total vol/ Vol per tank)		Tanks			
No. of frac pumps (est. from above)		Pumps			
Frac tank manifold		Units			
Hydration unit		Units			
No. of blenders		Units			
HI-LO pressure manifold		Units			
Frac pit manifold		Units			
High pressure manifold		Units			
Frac control van		Units			
Mobile lab		Units			
Crane truck and high pressure steel		Units			
Sand System1..		Units			
Sand System2..		Units			
Sand conveyor		Units			
Frac valve.................inch......................		Units			
Tree Saver.................inch......................		Units			
Pop-off valve.......inch..........psi..............		Units			
Check valves.............inch.....................		Units			
Plug valves.................inch.....................		Units			
Crew truck................Type.....................		Units			
Personnel		Persons			
Proppant 1............................Mesh............		LBS			
Proppant 2............................Mesh............		LBS			
Proppant 3............................Mesh............		LBS			
Gelling Agent1..		GAL			
Gelling Agent2..		GAL			
Stabilizer1..		GAL			
Stabilizer2..		GAL			
Breaker1..		GAL			
Breaker2..		GAL			
Crosslinker1..		GAL			
Crosslinker2..		GAL			
Crosslink Modifier1...............................		GAL			
Crosslink Modifier2...............................		GAL			
Buffer..		GAL			
pH Modifier..		GAL			
Bacteriacide..		GAL			
Non-emulsifier		GAL			
Surfactant..		GAL			
Friction reducer..		GAL			
Corrosion control..		GAL			
Scale control..		GAL			
Paraffin control..		GAL			
Clay stabilizer..		GAL			
Acid1.................Strength........................		GAL			
Acid2.................Strength........................		GAL			
Other Chem1..		GAL			
Other Chem2..		GAL			
Other Chem3..		GAL			

UFD	FLUID BLENDING AND QUALITY CONTROL

DATE	CUSTOMER
FIELD	WELL

PRODUCT NAME	AMOUNT	TRUCKING #

FRAC FLUID/SLURRY PROPERTIES		
	IDEAL	MEASURED
SPECIFIC GRAVITY		
VISCOSITY @ 300 RPM		
FLUID DENSITY		

IDEAL AND OBSERVED VISCOSITIES FOR DIFFERENT POLYMER LOADINGS

	POLYMER LOADING				
	A	B	C	D	E
TEMPERATURE					
IDEAL VISCOSITY (3 min)					
OBSERVED VISCOSITY (3 min)					
IDEAL VISCOSITY (10 min)					
OBSERVED VISCOSITY (10 min)					

BREAKER TESTS					
BREAK TIME @ TEMP	/	/	/	/	/

Comments

UFD PROPPANT SIEVE ANALYSIS AND QUALITY CONTROL

DATE			CUSTOMER			
FIELD			WELL			

TRUCKING NUMBER	PROPPANT TYPE	PROPPANT SIZE	% ABOVE	% BELOW	% FINES	ACCEPTABLE?

UFD

Comments

UFD WATER QUALITY CONTROL

DATE		CUSTOMER						
FIELD		WELL						

TANK #	1	2	3	4	5	6	7	8
pH (INITIAL)								
TEMPERATURE								
SPECIFIC GRAVITY								
CLARITY/ODOR								
CHLORIDES								
TDS								
PPM IRON								
ADDED CHEMICALS								
REDUSING AGENT								
BACTERIACIDE								
pH (FINAL)								

Comments

UFD

UFD ACID QUALITY CONTROL

DATE		CUSTOMER			
FIELD		WELL			

	TANK # 1	TANK # 2	TANK # 3	TANK # 4	TANK # 5
TRUCKING NUMBER					
TYPE OF ACID					
STRENGTH					
VOLUME					
TEMPERATURE					
SPECIFIC GRAVITY					
SALT (Y/N?)					
SALT TYPE					
SALT CONC.					
SOLVENTS (Y/N?)					
SOLVENT TYPE					
SOLVENT CONC.					
TO BE DILUTED (Y/N?)					
DILUTED STRENGTH					
DILUTION FLUID					
DILUTION FLUID VOLUME					
ACTUAL DILUTED STRENGTH					
ACTUAL DILUTED TEMP					
ACTUAL DILUTED SP. GR.					
(RETAIN SAMPLE OF ACID)					

UFD   **Calculations**

Sample Fracture Program

G

EXAMPLE FRACTURE WELL NO. B-4
ORCUTT FIELD, CA

OBJECTIVE: CONDUCT MINIFRAC AND FRACTURE TREAT THE FOXEN DIATOMITE

EQUIPMENT REQUIRED:

Pumping Equipment

- 7 : 500 BBL clean and inspected frac tanks
- 6 : V-12 PUMPS to deliver 1,000 HHP at 25 BPM (includes 50% backup)
- 1 : Blender
- 1 : LCG Pre-blender
- 1 : Chemical delivery truck
- 2 : Mountain Movers
- 1 : Tech Command Center

Special Equipment for Minifrac

- ■ "606" downhole pressure sensor/temperature probe and wireline service to provide continuous output for surface recording (Logging Services Co.)

- ■ Flowback manifold equipped with digital flowrate meter

- ■ 0–1000 psi and 0–5,000 psi transducers to monitor wellhead treating pressures

- ■ All necessary hardware and software for real-time recording and display of injection rate, flowback rate, wellhead treating pressure, downhole pressure, downhole temperature, and line tensions; all data will be played back in ASCII format and downloaded to Company computer on location

CLEAN FLUID VOLUMES REQUIRED:

Minifrac: 625 BBL
Main Frac: 2,467 BBL

PROPPANT REQUIRED:

314,400 LBS OF 20/40 MESH OTTOWA

. . . after well has been perforated at 725–925' and packer set at 675' on 3-1/2" tubing; assumes no rig on well . . .

SETUP

1. THE WEEK BEFORE THE MINIFRAC, MI AND SPOT 7 CLEANED AND INSPECTED 500 BBL FRAC TANKS. TREAT EACH TANK WITH 0.1% BE-3 & FILL WITH FRESH WATER ACCORDING TO THE ATTACHED *TANK MIXING SCHEDULE*.

   QUALITY CONTROL TESTING: Obtain water samples from each tank and record the tank number. For each sample, measure and record temperature, pH, and bacteria level. Gel samples from tanks 2–7 with 1% LGC-V according to the attached *Tank Mixing Schedule*. Measure and record the base gel viscosity for each sample. Determine and record the required crosslinker concentration.

2. THE DAY BEFORE THE MINIFRAC:

A. MI, spot, and RU fracturing company and equipment. Pressure test surface lines against the wellhead to 3,000 psi.

B. Pre-mix all tanks with 3% KCL. Pre-mix Tank 4 with Versagel 40. Measure and record the temperature, pH, and base gel viscosity of a Tank 4 sample after mixing. Perform a vortex closure test on a crosslinked sample to ensure adequate viscosity at downhole conditions.

C. RU Logging Service Co. with mast truck and 3,000 psi working pressure lubricator.

MINIFRAC

3. RIH W/ "606" PRESSURE GAUGE/TEMPERATURE PROBE. RUN BASE TEMPERATURE LOG FROM ED TO 300'. HANG TOP OF TOOL 10' BELOW BOTTOM PERFORATION. RECORD DEPTH OF PRESSURE SENSOR ON MORNING REPORT. CHECK CALIBRATION OF PRESSURE GAUGE.

4. ENSURE SURFACE EQUIPMENT IS READY TO START MINIFRAC.

A. Check data acquisition system to ensure injection meters, flowback meter, and wellhead pressure transducers are functioning properly.

B. Calibrate flowback controls and fill all lines with 3% KCL.

C. Pressure test all lines to 3,000 psi.

5. PERFORM MINIFRAC:

SPECIAL NOTE: **THE FOLLOWING SEQUENCE OF PUMPING AND MONITORING IS ONLY A GUIDELINE BASED ON ASSUMPTIONS. THE VOLUMES AND RATES WILL BE CHANGED BASED ON ACTUAL CONDITIONS ENCOUNTERED DURING THE TEST. ENGINEERING WILL BE ON LOCATION DURING THE MINIFRAC TO ADVISE ON PROCEDURE AS NECESSARY.**

Stage A **Filtration Test:** Pump 10 BBLS of 3% KCL water at 1 BPM. Hold wellhead pressure below 100 psi to avoid initiating fracture. Shut down and record pressure decline for 5 minutes.

Breakdown: Pump 50 BBLS of 3% KCL water at 10 BPM. Shut-in and monitor pressure decline for at least 30 minutes. Estimate breakdown pressure, ISIP, fracture propagation pressure, frac gradient, leak-off coefficient, and effective number of perforations open.

CAUTION: Maximum allowable surface pressure is 2,950 psi.

NOTE: IF AT ANY TIME DURING SHUT-IN THE WELLHEAD PRESSURE FALLS BELOW HYDROSTATIC ACCORDING TO THE 0–1,000 PSI TRANSDUCER, OPEN A BLEED VALVE AT THE WELLHEAD TO PREVENT VACUUM PRESSURE IN THE WELL.

Stage B **Extended Shut-in #1:** Re-open hydraulic fracture by pumping 3% KCL at 1 BPM. Once fracture has re-opened, increase pump rate to 10 BPM and pump 75 BBLS. Shut-in and monitor pressure until fracture closure is observed. Bound ISIP, leak-off coefficient, and minimum in-situ stress.

Stage C **Extended Shut-in #2:** Re-open hydraulic fracture by pumping 3% KCL at 1–3 BPM. Once fracture has re-opened, increase pump rate to 10 BPM and pump 100 BBLS. Shut-in and monitor pressure until fracture closure is observed. Estimate ISIP, leak-off coefficient, and minimum in-situ stress.

Stage D **Flowback Test:** Re-open hydraulic fracture by pumping 3% KCL at 1–3 BPM. Once fracture has re-opened, increase pump rate to 15 BPM and pump 150 BBLS. During the last 2 minutes, increase pump rate to 16 BPM and open flowback manifold to pit at 1 BPM. Shut down pumps and flow well back to pit at 1 BPM until fracture closure is observed. Confirm ISIP, leak-off coefficient, and minimum in-situ stress.

Stage E   **Gel Leak-off Test:** Re-open hydraulic fracture by pumping 3% KCL at 1–3 BPM. Once fracture has re-opened, switch to crosslinked Versagel 40 and increase pump rate to 25 BPM. Pump 250 BBLS of crosslinked frac gel and displace with 7 BBLS of 3% KCL. (Slow pump rate while pumping 3% KCL for displacement volume control.) Shut-in and monitor pressure until fracture closure is observed. Estimate gel leak-off coefficient to be used in design of the main fracture treatment.

6. RUN POST FRAC TEMPERATURE LOG FROM ED TO 300'. POOH W/ "606" PRESSURE GAUGE/TEMPERATURE PROBE. RD FRACTURING CO. LOGGING SERVICES. SECURE WELL.

7. IN PREPARATION FOR MAIN FRACTURE TREATMENT, RE-FILL TANKS 1 AND 4 WITH BACTERIACIDE TREATED 3% KCL WATER. PRE-MIX TANKS 4–7 WITH VERSAGEL 40 ACCORDING TO ATTACHED *TANK MIXING SCHEDULE.* MEASURE AND RECORD THE PH, TEMPERATURE, AND BASE GEL VISCOSITY FOR ALL GEL TANKS AFTER MIXING. PERFORM A VORTEX CLOSURE TEST ON A GELLED SAMPLE FROM EACH TANK TO ENSURE ADEQUATE VISCOSITY AT DOWNHOLE CONDITONS.

NOTE: Tanks 2 and 3 may also be pre-gelled depending on results of minifrac.

MAIN FRACTURE TREATMENT

8. ON THE DAY OF THE JOB, RE-TEST ALL LINES TO 3,000 PSI AND THOROUGHLY ROLL ALL PRE-MIXED TANKS. REPEAT GEL PH, TEMPERATURE, VISCOSITY, AND VORTEX CLOSURE MEASUREMENTS.

9. HYDRAULICALLY FRACTURE TREAT WELL B-4. REFER TO ATTACHED <u>TREATMENT SCHEDULE</u> FOR STAGE VOLUMES, GEL LOADINGS, AND PUMP RATES. **DO NOT OVERDISPLACE AND DO NOT BLEED OFF PRESSURE AFTER TREATMENT.**

CAUTION: Maximum allowable surface pressure is 2,950 psi

NOTE: ATTACHED *TREATMENT SCHEDULE* IS ONLY TENTATIVE; A REVISED SCHEDULE WILL BE GENERATED BY THE ENGINEER ON LOCATION BASED ON RESULTS OF THE MINIFRAC.

10. RECORD FINAL ISIP AND MONITOR SHUT-IN PRESSURE DECLINE FOR 1 HOUR OR UNTIL DIRECTED BY COMPANY ENGINEER ON LOCATION TO STOP MONITORING.

11. RD FRACTURING CO. LEAVE WELL SHUT-IN FOR A MINIMUM OF 48 HOURS.

    . . . MIRU workover rig. Clean well out. Note depth of fill and type on morning report. Complete well for production . . .

Table I. Tank Mixing Schedule

Tank	Base Fluid	Stages	Additives Batch Mixed	Additives On-the-Fly
		Minifrac		
1	3% KCL	A-D DISPL	0.1% BE-3 3% KCL	NONE
4	VERSAGEL 40	E	0.1% BE-3 3% KCL 1% LGC-V 0.05% MVF-2L	0.1% MVF-10 0.1 LB GBW-30
		Main Frac		
1	3% KCL	PRE-PAD FLUSH	0.1% BE-3 3% KCL	NONE
2,3	VERSAGEL 40	PAD	0.1% BE-3 3% KCL	1% LGC-V 0.05% MVF-2L 0.1% MVF-10 0.1 LB GBW-30
4,5,6,7	VERSAGEL 40	4–8 # SLF	0.1% BE-3 3% KCL 1% LGC-V 0.05% MVF-2L	0.1% MVF-10 0.1 LB GBW-30

Fracturing Co. Chemical Identification

LGC-V .... VERSAGEL LT
MYF-2L .... PH BUFFER
MYF-10 .... CROSSLINKER (ANTIMONY)
GBW-30 .... BREAKER
BE-3 .... BACTERIACIDE

NOTE: TANKS 2 AND 3 MAY NOT BE REQUIRED DEPENDING ON RESULTS OF MINIFRAC; THESE TANKS SHOULD NOT BE GELLED AHEAD OF TIME.

Table II. Treatment Schedule (tentative)

	Stage	Clean Fluid (bbl)	Stage Description	Sand Vol (lbs)	Slurry Vol (bbl)	Pump Rate (bpm)	Pump Time (mins)
	A	50	Breakdown – KCL	—	50	10	5
	B	75	Extended SI – KCL	—	75	10	7.5
Minifrac	C	100	Extended SI – KCL	—	100	10	10
	D	150	Flowback – KCL	—	150	15	10
	E	250	Extended SI – Gel	—	250	25	10
			(Displace w/ KCL)				
	Total	625					
	1	50	Pre-Pad – KCL	—	50	25	2.0
	2	—	ISIP	—	—	25	—
	3	695	Pad – Gel	—	695	25	27.8
Main Frac	4	169	SLF – 2 ppg	14,196	184	25	7.4
	5	263	SLF – 3 ppg	33,151	299	25	12.0
	6	392	SLF – 4 ppg	65,880	463	25	18.5
	7	560	SLF – 5 ppg	117,643	687	25	27.5
	8	331	SLF – 6 ppg	83,530	421	25	16.8
	9	7	Flush – KCL	—	7	10	0.7
	Total	2,467		314,400			112.7

Index

A

anisotropy, 196
areal proppant concentration, 10,
  204–205

B

bilinear flow, 191
Bingham plastic, 44
biocides, 85
Biot's constant, 90–91
Blender, 168
breakers, 86,87

C

CDM (*see* continuum damage
  mechanics)
candidate reservoirs, 2, 3, 6, 12, 58
carbon dioxide foam, 84
chemical mixing, 168
clay control additives, 86
closure pressure (stress), 66, 88–91,
  102, 124, 125

communications, 173
complex wells, 13
connectivity (fracture-to-well), 8
continuum damage mechanics,
  134–135
crosslinked fluids, 43, 94

D

damage
  choke, 95
  fracture face, 95
  reduction in proppant pack
    permeability, 95
deviated wells, 9, 13
design optimization, 5, 6, 7–8, 29,
  74–76
dilatancy, 134

E

efficiency, 47, 106
ellipsoid flow, 45
embedment, 7
  design example, 145–148

encapsulated breakers, 87
enzymes, 87
equipment, fracturing, 164–174
Euler gamma function, 49

F

filter cake, 68, 69, 94
flow behavior index, 44
fluid lag, 134
fluid loss control additives, 85
fluids
  additives, 83, 85–87
  fluid mechanics, 42–45
  selection, 94–95
foaming agents, 86
frac van, 172
Forchheimer equation, 77
frac&pack (*see* high permeability
  fracturing)
fracture conductivity, 6, 9, 27, 34,
  35–37, 61, 77, 79, 80, 87, 88,
  91–99
fracture design, (*see also* HF2D),
  3, 4, 7–8, 14–15, 37–38,
  101–127
fracture compliance, 104
fracture dimensions
  azimuth, 9
  length, 5, 6, 7, 35, 36, 38,
    92–93
  height, 7, 107, 109, 111,
    129–134, 186–187
  penetration ratio, 27
  width (hydraulic), 41, 47, 51–55,
    92–93, 106, 110
  width (non-Newtonian), 113
  width (propped), 5, 6, 7, 10, 36,
    38, 118
  width (soft formations), 66
fracture propagation, 11, 66,
  134–135
fracture stiffness, 104
fracture toughness, apparent, 134
fracturing fluid penetration, 98

fracturing fluids
  borate fluids, 84–85
  crosslinked, 43, 59, 84–85
  emulsions, 84
  foams, 84
  linear gells, 59
  titanate fluids, 84–85
  zirconate fluids, 84–85
friction pressure, 124
friction reducers, 86
forced closure, 181–182

G

gas condensate reservoirs, 72–76
gravel pack, 60

H

HF2D, 15, 119
HF2D, medium permeability
  example, 137–142
HF2D, "pushing the limits
  example," 142–145
HF2D, high permeability example,
  148–152
HF2D, extreme high permeability
  example, 152–157
HF2D, low permeability example,
  157–162
HPF (*see* high permeability
  fracturing)
HPG (*see* hydroxypropyl guar)
HI-LO pressure manifold, 169
height, 7, 107, 109, 111
  containment, 186–187
  design for, 129–134
high permeability fracturing, 1, 10,
  13, 57–82, 93–99, 119–122
  evaluation, 193–206
high-rate water packs, 62
high-temperature oxidizers, 87
history of fracturing, 1, 57
hook-up, 174–180
horizontal wells, 9, 13, 62

hydration unit, 168
hydroxypropyl guar, 85
hypergeometric function, 104

I

ISIP (*see* instantaneous shut-in
    pressure)
injected volume calculation, 110
injection tests
    microfracture, 101
    minifracs, 102–109
instantaneous shut-in pressure,
    102–103

K

KGD geometry, 41, 49, 53–54, 92,
    105, 111
Kachanov parameter, 134

L

LEFM (*see* linear elasticity)
leakoff, 12, 46–50, 67–72
leakoff coefficient, 46, 68,
    102–109, 194
linear elasticity, 39–42, 134
linear flow, 192
logging methods, 188–189

M

mapping, 189–190
material balance, 46–48, 110
microfracture, 101
minifrac, 102–109, 125–126
monitoring, 179–180,

N

net present value (NPV), 8
net pressure, 11, 41, 51, 66, 131
Newtonian fluid, 43, 44
nitrogen foam, 84
Nolte analysis, 102–109
Nolte-Smith analysis, 185–186

non-Darcy effects, 7, 32, 76–82,
    135–136
opening time distribution factor, 46,
    47, 111

P

PKN geometry, 41, 49, 51–53, 92,
    105, 111
packing radius, 198, 203
pad, 113, 116
perforating, 3, 9, 122, 124
persulfates, 87
pit manifold, 166
plane strain modulus, 40
plastic fluid, 43
Poisson ratio, 39–40, 90
polymer invaded zone, 70–72
poroelasticity, 90–91
Power law (fluid), 44
Power law (fracture propagation),
    48–49, 67
pressure falloff tests, 126
pressure measurements
    bottomhole, 126
    deadstring pressure, 127
    washpipe data, 127
pre-treatment tests for HPF,
    122–127
    step-rate tests, 123–125
    minifracs, 125–126
    pressure falloff tests, 126
productivity index, 5, 23–26, 60,
    75, 136
proppant number, 27–32, 74, 109,
    129
proppant schedule, 113–118
proppant slurry, 113–118
proppants, 5, 83, 87–91
    areal concentration, 10, 204–205
    bauxite, 87
    ceramic, 87
    distribution in fracture, 113–115
    embedment, 7
    sands, 87

Proppants *(continued)*
  selection guide, 89, 93–94
  strength, 87
  volumetric efficiency, 7, 37
pseudo-3D models, 133
pseudosteady state flow, 24
pseudoplastic fluid, 43
pump time, 110–113
pumps, 170

Q

quality control, 163, 173, 180–181,
    183

R

radial fracture, 54–55, 105
real-time analysis, 185–186
real-time pressure data, 194–195
relative permeability effects, 73

S

sand control, 58
sand conveyor, 167
sand supply system, 167
seismic imaging, 189
shape factor, 52, 53
shear modulus, 40
skin
  composite, 95–98
  damage skin, 5, 24
  fracture, 25, 33, 61, 82
  fracture face, 73–74, 136–137,
    196
  gravel pack, 60
  for HPF, 197
  mechanical, 25
  non-Darcy, 79, 80, 82
  radial equivalent, 4
slopes analysis, 197
slot flow, 45
spurt loss, 46, 68, 105
steady state flow, 24
stress intensity factor, 42

stresses
  absolute, 90
  effective, 90
  principal, 8, 90
  values, 39, 41
surfactants, 86

T

tanks, 167
tilt meters, 189–190
tracers, 188
transient flow, 24
TSO (*see* tip screenout)
tip screenout, 6, 9–11, 57, 63–65,
    92, 115
tip screenout design, 119–122
  design examples, 148–157
tip effects, 134–135
tortuosity, 9
transfer pump, 166
two-dimensional (2D) models,
    50–55

V

viscosity
  apparent, 43, 44
  effective, 71
  Newtonian, 44
  polymer solution, 85
viscoelastic (VES) fluids, 99

W

well deviation, 3
well performance, 5–7, 32–35
wellbore radius, equivalent, 33
well testing, 190–193
  diagnostic plot, 191
  specialized plot, 192
  post-treatment for HPF, 195–197

Y

Young's modulus, 39–40